FUNDED BY

Culture Development and Promotion Center of
Haidian District, Beijing

BUU "Beijing Studies"
A Top Discipline In Beijing Universities

Honors to Professor Meng Zhaozhen, Mr. Zhang Baozhang and all experts and leaders who have offered us great support and encouragement.

Thanks to the School of Landscape Architecture of Beijing Forestry University for years of cultivation.

Thanks to the Culture Development and Promotion Center of Haidian District and
Beijing Union University for their kind sponsor for this book.

Thanks to Tang Lu, Zhong Yuan, Qian Yun and other experts for reviewing the manuscript.

- 
- 
- 
- 
- 

Research Team of the Three Hills and Five Gardens of Beijing Forestry University

景致 | viewpoints

— Longevity Hill

— Jade Spring Hills

— Fragrant Hills



— Yuanming Yuan

— Changchun Yuan

— Qingyi Yuan

— Jingming Yuan

— Jingyi Yuan

# Touring the Three Hills and Five Gardens

## Landscape, Art and Life of Chinese Imperial Gardens

**Author** Zhu Qiang et al.
**Translator** She Sha

China Forestry Publishing House

**MEMBERS OF THE AUTHOR TEAM**
Zhu Qiang
Wang Yize / Cong Xin / Fu Jing /
Song Qi / Cao Shuyi / Du Yican / Guo Jia

图书在版编目（CIP）数据

三山五园 = Touring the Three Hills and Five Gardens——Landscape, Art and Life of Chinese Imperial Gardens : 英文 / 朱强等著; 佘莎译. — 北京: 中国林业出版社, 2020.11

ISBN 978-7-5219-0791-9

Ⅰ.①三… Ⅱ.①朱… ②佘… Ⅲ.①古典园林—介绍—北京—英文 Ⅳ.①K928.73

中国版本图书馆CIP数据核字(2020)第175146号

Executive Editor: Sun Yao
Cover Illustrator: Liu Mumao

Published by China Forestry Publishing House,
7, Liuhai Hutong, West City District, Beijing, P.R.China, 100009
http://www.forestry.gov.cn/lycb.html
Copyright © 2019 by China Forestry Publishing House.
All rights reserved. Unless permission is granted, this material shall not be copied, reproduced, or coded for reproduction by any electrical, mechanical, or chemical process or combination thereof, now known or later developed.
Printed the People' s Republic of China.
Archives Library to Chinese Publications Number 2020-175146
International Standard Book Number   978-7-5219-0791-9
Touring the Three Hills and Five Gardens——Landscape, Art and Life of Chinese Imperial Gardens / Author: Zhu Qiang et al. / Translator: She Sha, 2020.9
I. ① Three... II. ① Zhu...III. Imperial Gardens - Introduce - Beijing- English IV.①K928.73

Tel: 010-83143629
E-mail:JZ_view@163.com
Price: CNY 148.00

修凿深泉便汲

朱强等同道共勉

己亥大暑

孟兆植

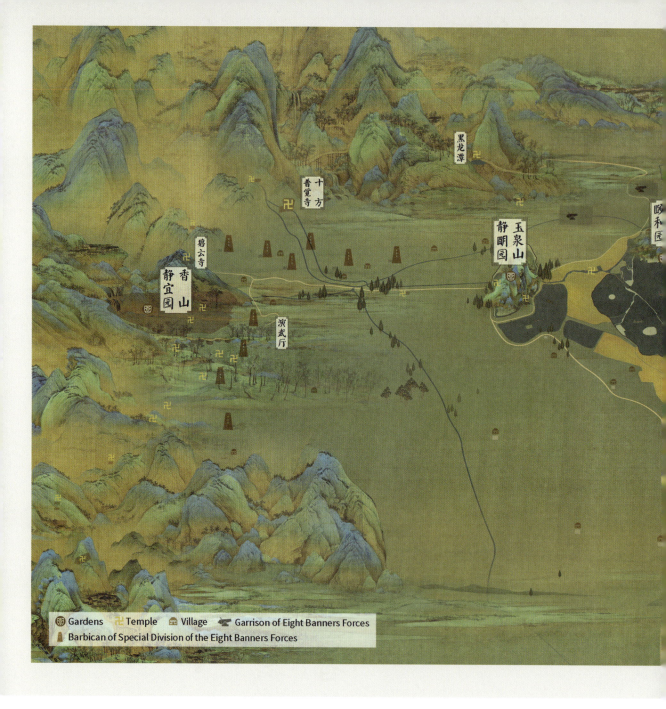

## Restored Panorama of the Three Hills and Five Gardens (abbr. THFG) Region in Its Heyday

This birdview map shows a vivid picture of the gorgeous sight of the THFG region in an archaistic fashion. On the browned scroll, turquoise green hills undulate one upon another; dark blue waters stretch far into the distance; dozens of gardens shine off their

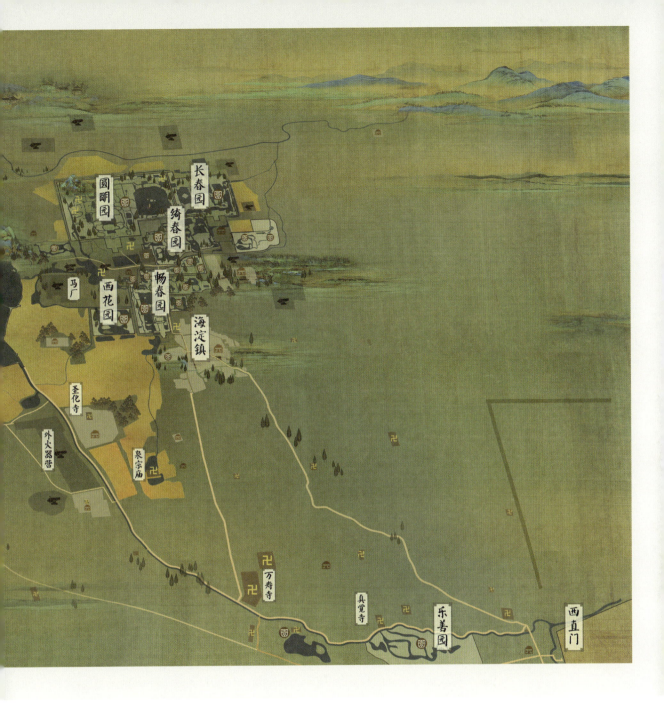

magnificence against the background of golden-colored paddy fields across the western suburb Haidian. This best places in the capital, as Emperor Kangxi acclaimed in his notes, presents itself like a nature-made picture. Temples, villages and military barracks are dotted here and there. The painting combines a famous Northern Song art piece, *A Panorama of Rivers and Mountains*, with contemporary restorations of the ruins through digital techniques. In addition to causing the hills and waters to resemble the Western Hills, Jade Spring Hills and Longevity Hill, it also places trees and boats at the right points, adding life to the entire picture.

# FOREWORD

A young friend of mine, PhD student Zhu Qiang from Beijing Forestry University, handed to me the manuscript of a book, *Touring the Three Hills and Five Gardens—Landscape, Art and Life of Chinese Imperial Gardens*. With great interest I went through the manuscript and said to myself: This is a nice book well worth reading. It's a vivid picture of the cluster of imperial gardens in Haidian.

I always maintain that water is the lifeline of the Three Hills and Five Gardens. Here in this book, a large portion is used to describe how water was harnessed and used in garden-making. First, the book records how Emperor Qianlong, guided by the "hydro-based development" philosophy, utilized the rich water resources at the Western Hills through 4 steps. He expanded the Urn Hill Lake, dredged up the Long River, broadened the High-Level Water Lake and the Water Storage Lake and installed the aqueducts at the Fragrant Hills; he built each of the imperial gardens in the western suburbs into a waterscape wonderland; he designed and constructed a number of famous water features in each of the gardens. The book also includes a detailed account of the routes the emperor would take to tour the gardens along the imperial waterway, and how he viewed and praised the waterscape there. In fact, water features are highlighted in each description of the THFG: the chapter title of "Garden construction originally intended for water conservancy""Imperial palaces incorporating the charm of woods and springs" and "The heart of the empire amid artificial hills and waters". From these descriptions a conclusion is drawn: Water systems are the blood veins of the THFG region. They are just a few examples showing the authors' deep understanding and wonderful presentation of this proposition.

In my point of view, imperial characters are the soul of the THFG. The book elaborates on the vastness of the gardens, the sumptuousness of the buildings in them, and the suitability of the gardens for life and work. It gives a live illustration of how the emperors showed up in the imperial court and dealt with state affairs. It also includes monographic descriptions of the intricate, secret imperial family lives, such as the harem routines, festive celebrations, and riding and arrow-shooting drills. These fully demonstrate that the imperial gardens in Haidian served as a second national political center besides the Forbidden City.

To me, the profound connotation of the traditional Chinese culture is the greatest historical value of the THFG. The book reveals an undeniable fact: the design of the gardens reflects the ruling policy of the Qing emperors. That is, they were more inclined toward agriculture and Tibetan Buddhism. The arrangement of landscapes, the placement of artificial hills and waters, and the naming of buildings all conveyed an interpretation of traditional cultural ideas and aesthetic perceptions throughout. Renowned garden scenes from the regions south of the Yangtze River and other parts of the country were creatively transplanted into the imperial pleasure—this not only makes the imperial gardens an assembly of the greatest garden landscapes in the country, but also "showcases the openness and inclusiveness of the THFG".

Backed by a great diversity of historical texts, the book portrays a unique, picturesque image of the imperial gardens. The success of this book lies in the leading author Zhu Qiang and his THFG research team from Beijing Forestry University. This vibrant, young group of garden theory explorers, with their diligence and readiness for innovation, set a challenging but very meaningful goal for the book: to launch a restoration research and popularization campaign on the THFG with professional knowledge. The publication of this book is a record of their success. From the bottom of my heart, I congratulate them on their success and admire the great contribution they have made. In the afterword, they say, "My participation in compiling this book is an unforgettable experience in my life. It also marks the beginning of my story with the THFG. I hope I will continue to contribute to the THFG together with the other members of the team." I was very pleased at these words. We hope that more young elites with ideals and ambitions will join in the research and popularization of the THFG like Zhu Qiang and his team. Together, let us polish this "golden name card" of Haidian and Beijing and build a perfect THFG historical cultural scenic area.

**Note**

Zhang Baozhang, Beijing history-geography folk customs and imperial gardens expert, one of the reviewers of the book.

July, 2019

# INTRODUCTION

Imperial gardens were important part of the imperial life in ancient China. They are also an important category of Chinese gardens. Lingyou of Zhou dynasty, Epang Palace of Qin dynasty, Shanglin Palace of Han dynasty, Daming Palace of Tang dynasty, Genyue of Song dynasty, and Taiye Pond of Yuan dynasty, etc., are all the crowning art achievements of their respective dynasties. There, episodes of historical romances have been staged one after another.

In the Qing dynasty, the last feudal dynasty in the Chinese history (1644—1911), the emperors were not confined to residing in the Forbidden City compound. Instead, 7 generations of the emperors, including Shunzhi, Kangxi, Yongzheng, Qianlong, Jiaqing, Daoguang and Xianfeng, chose to continuously build luxurious and comfortable palaces on a beautiful and fertile plot of land in the western suburbs of Beijing. For this purpose, the most skillful craftsmen in the country and huge sums of imperial funds were mobilized. The generations of the Qing emperors created 5 large imperial gardens across the 3 famous hills and the vast tracts of wetland around them. Each of them spent at least a half of the year in average working and residing there. The gardens were densely surrounded by government institutions, smaller gardens granted by the emperors to their clans or ministers, religious temples, imperial guard barracks, water conservancy facilities, farmlands and villages. These elements compound to make up the THFG, an imperial political center on the outskirts of the capital. At its prime, this garden cluster covered a larger land area even than Beijing. That is precisely the focus of concern in this book.

The term "THFG" outlines the core elements of this area—the imperial gardens. More exactly, it refers to the three hills and the three gardens built on them—the Fragrant Hills cum Jingyi Yuan, the Jade Spring Hills cum Jingming Yuan and the Longevity Hill cum Yihe Yuan (also known as Summer Palace, named Qingyi Yuan before Guangxu period)—which were all travel palaces of the emperors for worship and pleasure, plus two other gardens, Changchun Yuan and Yuanming Yuan (also known as the Old Summer Palace). They were detached palaces inhabited by the imperial family most of the year. According to statistics, in the middle to late Qing dynasty (the first half of the 19th century), the THFG region at least included 10 imperial gardens and 14 granted gardens. The total land area was as large as 1,175.6 hm². Imperial gardens covered 85.8% of the area. Their total area was approximately 14 times the size of the Forbidden City. Because of their extraordinary artistic taste, in the 18th to 19th century, they were brought to fame in Europe by missionaries' letters and paintings, making "Chinoiserie" a symbol for the aesthetic vogue of that time.

Frustratingly, the decline of feudal dynasticism exposed the THFG to severe devastation by the western invaders in 1860. The Anglo-French Allied Forces, although composed of less than 20 thousand men, not only captured the capital of the Qing empire in a short time, but also looted the imperial gardens and set them on fire. This evil conduct is undoubtedly a black mark in the human history of civilization and a heavy loss for the oriental culture. For this reason, the great French man of letters Victor Hugo once voiced strong condemnation against the war. In the last years of the dynasty, the Qing monarchs tried hard to restore the gardens with the poor stock of budget available, only to prove their decline to be irreversible. Today, the entire area has taken on a completely new look. The imperial temple for offering sacrifice to Emperor Kangxi is overshadowed by the bustles and hurries of modern streets. The barbicans for drilling special ladder troops are standing by themselves in the wooded hills. The streams once meandering through the gardens are entrapped in cement pools.

Although more than one and a half centuries has elapsed, the culture and art contained in the THFG still glitters in a unique glory. Its rise and fall are inscribed in our brains, offering us profound lessons and inspirations.

When we open a design sketch drawn by ancient craftsmen, a garden scene depicted by court painters, an imperial verse written by the emperor, or a court file transcribed by the ministers, we will be amazed that the THFG is no longer a fragment of history or an empty heritage site. Nor is it a pure garden landscape by itself. It is an epitome of a real-life ancient society comprised of the imperial family, ministers, farmers, merchants, craftsmen, monks, and armymen. Here there are not only the venues for political diplomacy and military drills, but also the daily routines and artistic lives of the imperial family, as well as the painstaking elaboration of officials and the sweating labor of civilians... In fact, everyone that appeared there and everything that occurred there have helped make this cultural heritage more vivid and lively. The more we know about the THFG, the more firmly we believe that its lost splendor is well worth being known by more people.

The authors of this book are students majoring in the art of the traditional Chinese garden. With rigorous attitude, simple words and vivid pictures, they restored the creation process of the THFG and the lifestyle of the ancients in these gardens, and interpreted the underlying scientific, cultural and artistic values. Besides, the book makes rewarding exploration and attempt to revealing and disseminating traditional Chinese ecological ideas, cultural deposits, aesthetic orientations and lifestyles. By reading the book, readers would find that the imperial gardens and the imperial family lives are no longer as mysterious as you have imagined. You would be captivated by the cultural glamour of the THFG and even decide to view it for yourself, or begin to contemplate on the connections between tradition and modernity. Sincerely hope this book will be instructive and helpful to dear readers.

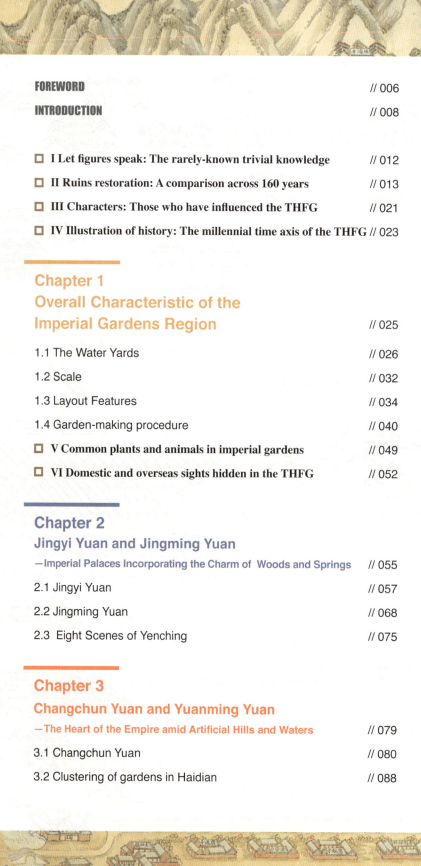

| | |
|---|---|
| **FOREWORD** | // 006 |
| **INTRODUCTION** | // 008 |

☐ I Let figures speak: The rarely-known trivial knowledge // 012
☐ II Ruins restoration: A comparison across 160 years // 013
☐ III Characters: Those who have influenced the THFG // 021
☐ IV Illustration of history: The millennial time axis of the THFG // 023

## Chapter 1
## Overall Characteristic of the Imperial Gardens Region // 025

1.1 The Water Yards // 026
1.2 Scale // 032
1.3 Layout Features // 034
1.4 Garden-making procedure // 040
☐ V Common plants and animals in imperial gardens // 049
☐ VI Domestic and overseas sights hidden in the THFG // 052

## Chapter 2
## Jingyi Yuan and Jingming Yuan
—Imperial Palaces Incorporating the Charm of Woods and Springs // 055

2.1 Jingyi Yuan // 057
2.2 Jingming Yuan // 068
2.3 Eight Scenes of Yenching // 075

## Chapter 3
## Changchun Yuan and Yuanming Yuan
—The Heart of the Empire amid Artificial Hills and Waters // 079

3.1 Changchun Yuan // 080
3.2 Clustering of gardens in Haidian // 088

3.3 Yuanming Yuan // 092

☐ **VII Yuanming Yuan in the eyes of a French missionary who worked for the emperor** // 105

3.4 Charngchun Yuan and European Palaces // 106

3.5 The darkest days // 114

## Chapter 4
## From Qingyi Yuan to Yihe Yuan
—Garden Construction Originally Intended for Water Conservancy // 123

4.1 Tracing back to the origin // 124

4.2 Qingyi Yuan // 131

4.3 Ready in shape // 142

4.4 From Qingyi to Yihe // 144

## Chapter 5
## Garden-based Life of the Imperial Family // 151

5.1 Court at front, life at rear // 153

5.2 Landscape tours // 162

☐ **VIII Calendar of the Qing imperial family in the THFG** // 172

5.3 Festival celebrations // 174

☐ **IX Lantern Festival agendas and menus at Yuanming Yuan** // 179

5.4 Military drill // 180

5.5 Administration and management // 183

**CONCLUSION** // 190

**GLOSSARY** // 192

**REFERENCES** // 199

**AFTERWORD** // 201

# Let figures speak: The rarely-known trivial knowledge

## 1. Gardens Having Outstanding Advantages Both in Quantity and Scale

During the reign of Emperor Xianfeng (the mid-19th century)

- **10** Imperial gardens
- **14** Granted gardens
- A total area of **1175.6 hm²**

**1007.1 hm²** Total area of imperial gardens
≈ **14** Forbidden City
Versailles Palace: 800 hm²

961 m × 753 m

## 2. Landscaping Cultural Traditions Passed From Generation to Generation

**The oldest imperial gardens:**
Spring Water Yard at Jade Spring Hills
Pond Water Yard at Fragrant Hills
Built in **1190**

**The largest imperial garden:**
Yuanming Yuan (Old Summer Palace)
(including Charngchun Yuan and Qichun Yuan)
Covering an area of **355** hm²

**114** theme scenic areas were inscribed by Emperor Qianlong and Emperor Jiaqing:
- Yuanming Yuan (Old Summer Palace) 40
- Jingyi Yuan 28
- Jingming Yuan 16
- Qichun Yuan 30

## 3. Praising the Landscape with Vast Number of Poems

Emperor Qianlong wrote **6530** poems for the Three Hills and Five Gardens:

- Changchun Yuan **127 poems**
- Yuanming Yuan **2300 poems**
- Fragrant Hills Jingyi Yuan **1480 poems**
- Jade Spring Hills Jingming Yuan **1100 poems**
- Longevity Hill Qingyi Yuan **1523 poems**

## 4. Spring Water Ceaselessly Flowing to the Capital City

During the reign of Emperor Qianlong
**35** spring openings
in Haidian and Jade Spring Hills

The paddy field irrigated by the spring water
**5370910.32** m²
Rental: **4** qian ~ **8.2** qian silver / mu
(1qian = 3.125g  1mu = 666.67m²)

## 5. Gardens Gained Special Favor From the Emperors

The average staying time of the emperors in Qing Dynasty (Day/Year):

| Kangxi (1662—1722) | Yongzheng (1723—1735) | Qianlong (1736—1795) | Jiaqing (1796—1820) | Daoguang (1821—1850) | Xianfeng (1851—1861) |
|---|---|---|---|---|---|
| 107.5 | 206.8 | 126.6 | 162.0 | 260.1 | 216.4 |

## 6. Managing and Protecting the Gardens with Heart and Soul

During the reign of Emperor Qianlong,
**1947** people in the management team of the Five Gardens

During the reign of Emperor Xianfeng,
**87781** sets of ornaments and furniture in the buildings of Three Hills
(78185 of them were lost or destroyed in the war of 1860)

During the reign of Emperor Xuantong,
**16478** soldiers and their family members in the Guard Division of the Eight Banners Forces of Yuanming Yuan

- Yellow Banner Force **1932**
- White Banner Force **2159**
- Red Banner Force **2150**
- Blue Banner Force **2120**
- Red-ringed Yellow Banner Force **1961**
- Red-ringed White Banner Force **2086**
- White-ringed Red Banner Force **1959**
- Red-ringed Blue Banner Force **1961**

# II

## Ruins restoration: A comparison across 160 years

Although we can see many historical relics still existing today, it is not exaggerating to say that they are merely a tip of an iceberg (Fig.I). The absolute majority of the ancient gardens and buildings, despite their extreme popularity at their times, are submerged in the long river of history. The cultural heritage of the THFG, for example, has not left behind much for us to see since the Anglo-French invasion of China 160 years ago. Not only have many gardens and buildings like Changchun Yuan disappeared for good, even in those existing ones—like Yuanming Yuan, Jingyi Yuan, and Jingming Yuan—there are numerous ruins. The water systems, paddy fields and villages that used to surround the gardens have almost all been restyled by city development. How shall we come to know its original, blooming panorama(Fig.II)? This is a question that all historians are asking themselves today.

Fortunately, we can solve this puzzle through RESTORATION RESEARCH. As an important part of the research of garden or architectural history, restoration research involves the use of all data carrying historical information: ancient court archives and paintings, and modern aerial photographs, survey maps and archaeological reports... Any documented record containing historical information can be utilized in the restoration effort. In the meantime, it also demands high competence of the researchers themselves. One has to "patch up" the visage of the elements contained in a site at different periods of times through intensive study and visualize them to the public in the form of restored plans. This complicated process of study may be highly time and energy consuming, yet it enables us to probe into the historical information behind the already non-existent relics and mine out more of its historical values. For this reason, restoration research is of great significance.

In the beginning, our focus was mainly on the hundred scenes inside the imperial gardens themselves. Later, as our study deepened, we came to realize that gardens were only part of the THFG system. Then, did it have an overall blueprint, like the City of Beijing in the Ming and Qing dynasties? What were the artistic features and construction processes behind each of these gardens? What were the activities of the Qing monarchs and other imperial family members inside these gardens? What were the management activities that enabled these beautiful scenes to sustain hundreds of years?... With these questions, we set off on a long expedition of exploration in 2015. Up to the present day, we have initially found an answer (Fig.III-Fig.IV).

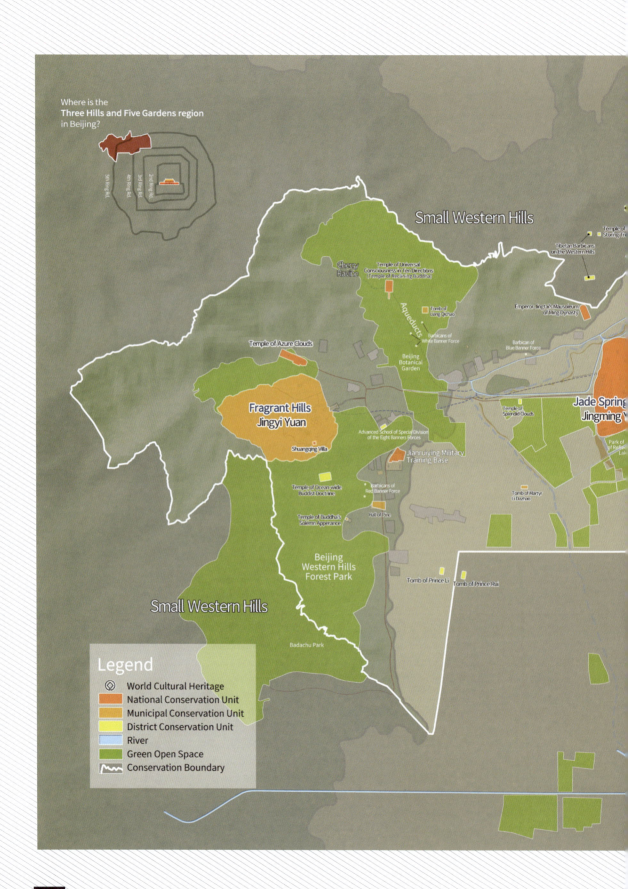

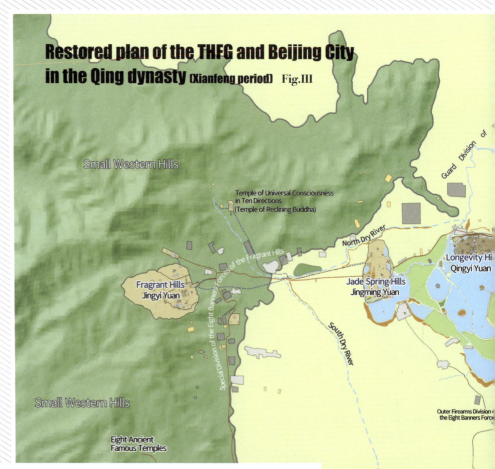

**Restored plan of the THFG and Beijing City in the Qing dynasty (Xianfeng period)** Fig.III

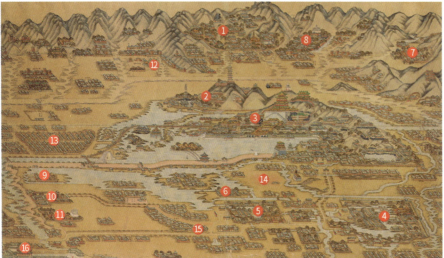

**Fig.II**
Qing Painter. *A Panorama of the Summer Palace and Surrounding Eight Banners Barracks* (U.S. Library of Congress)

1　Fragrant Hills cum Jingyi Yuan
2　Jade Spring Hills cum Jingming Yuan
3　Longevity Hill cum Yihe Yuan (Summer Palace)
4　Yuanming Yuan (Old Summer Palace)
5　Changchun Yuan
6　West Garden
7　Temple of Universal Consciousness in Ten Directions
8　Temple of Azure Clouds
9　Quanzong Miao
10　Longevity Temple
11　Temple of True Awakening
12　Hall of Reviewing Troops
13　Outer Firearms Division of the Eight Banners Forces
14　Imperial Racecourse
15　Haidian Town
16　Xizhi Gate

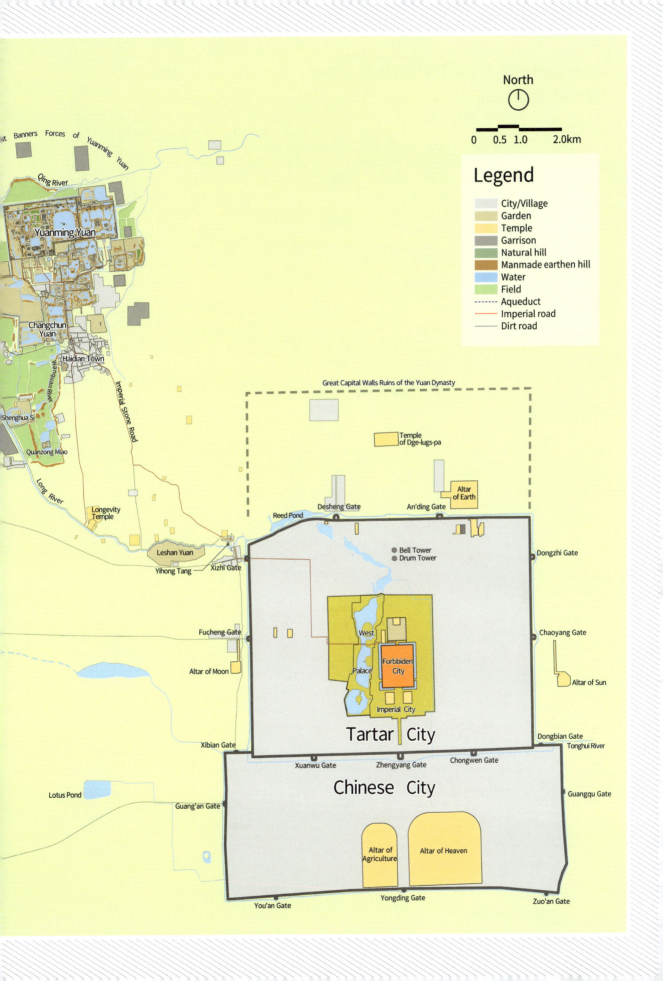

Restored plan of the THFG Region in the Qing dynasty (Xianfeng period) Fig.IV

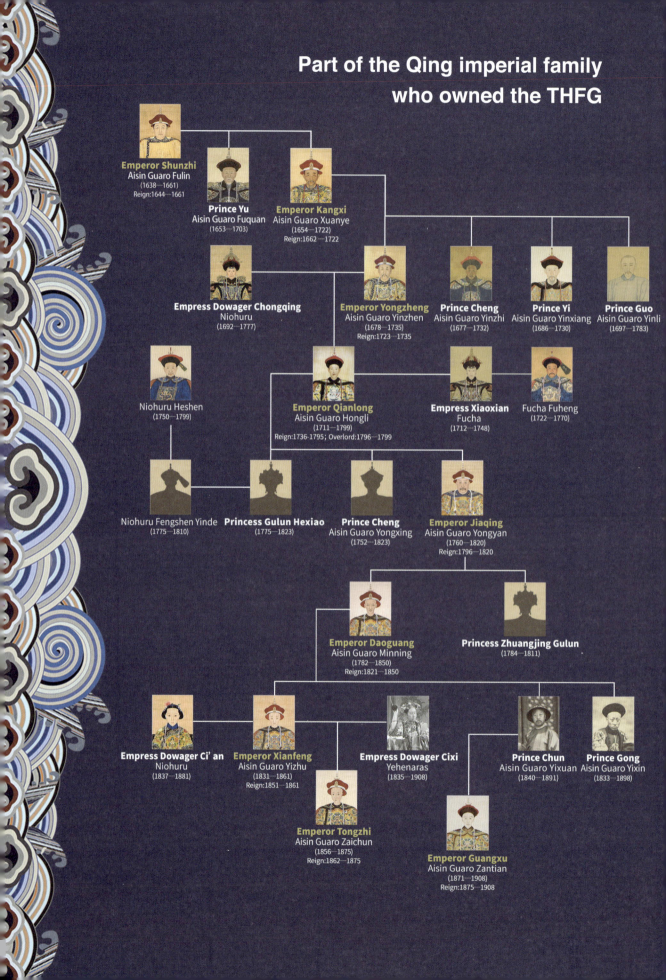

# III. Characters: Those who have influenced the THFG

## Garden-making pioneers

**Jin Emperor Zhangzong**
**Wanyan Jing (1168—1208)**
Regin: 1189—1208

6th emperor of the Jin dynasty. He built the "Eight Water Yards" of the Western Hills at the Fragrant Hills and Jade Spring Hills and visited there many times. He gave the name "Eight Scenes of Yenching".

## Religious leaders

**The 6th panchen lama**
**Lobzang Palden Yeshe (1738—1780)**

The highest-rank religious leader of Tibet. Emperor Qianlong built the Temple as Magnificent as Jokhang at Jingyi Yuan for him as his temporary palace when he was offering the birthday congratulations to the emperor in Beijing.

## Top court painters

**Tang Dai (1673—1752)**
**Shen Yuan (1736—1795)**

Painters who co-painted the famous *Album of Paintings of Forty Scenes of Yuanming Yuan*.

## European missionaries

**Lang Shining (1688—1766)**
Italian name: Giuseppe Castiglione

A famous court artist from Milan, Italy. Working in Yuanming Yuan, he was good at painting human portraits, plants and animals. His unique painting approach of combining Chinese and Western Painting expertise was very appealing to the emperors.

**Dong Bangda (1696—1769)**

Civil minister and painter. One of his representative piece was the *Scroll Painting of Jingyi Yuan*.

**Jiang Youren (1715—1774)**
French name: P. Benoist Michel

A Jesuit priest from France and famous court artist who participated in the design of European Palaces at Yuanming Yuan. His representative pieces include the *Great Universal Geographic Map*.

## Craftsmen (Leis Family)

**Lei Tingchang**

The Leis worked at the Style House, the imperial engineering department for 8 consecutive generations and served as chief engineer for many times. For this reason, they are known as "Yangshi Lei".

**1st generation**
Lei Fada
1619—1693
(projects involved:) Imperial Palace

**2nd generation**
Lei Jinyu
1659—1729
Changchun Yuan, Yuanming Yuan

**3rd generation**
Lei Shengcheng
1729—1792
Imperial engineering works during Qianlong period

**4th generation**
Lei Jiawei (1758—1845)
Lei Jiaxi (1764—1825)
Lei Jiarui (1770—1830)
Yuanming Yuan, Changchun Yuan, Qingyi Yuan, Jingming Yuan, Jingyi Yuan, etc.

**5th generation**
Lei Jingxiu
1803—1866
Yuanming Yuan, etc.

**6th generation**
Lei Siqi
1826—1876
Yuanming Yuan, West Palace, etc.

**7th generation**
Lei Tingchang
1845—1907
Yihe Yuan, Yuanming Yuan, West Palace, etc.

**8th generation**
Lei Xiancai
1877—?
Yihe Yuan, Yuanming Yuan, West Palace, etc.

## Invaders

**James Bruce, 8th earl of Elgin**
(1811—1863)

**Jean-Baptiste Louis Gros**
(1793—1870)

Special envoys of Great Britain and France who launched the Second Opium War. They coerced the Qing court into signing the *Tianjin Treaty* in 1858, and after failing to modify the treaty, jointly invaded China again in 1860, during which they looted Yuanming Yuan and the Three Hills; Elgin even ordered his British forces to burn down all the imperial gardens in Haidian. The two countries signed the *Beijing Treaty* with Qing government at the end of October.

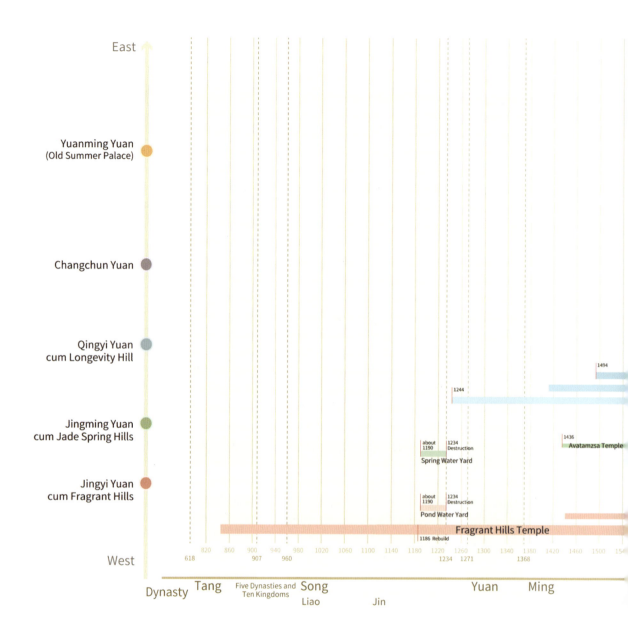

## How to read the illustration

From left to right, the coordinate represents the millennial historical vein from the Tang dynasty to the Qing dynasty. The times of reign of different Qing emperors are distinguished with dotted lines; the milestones of events within the same garden are denoted with text. The years 1860 and 1900, which witnessed the invaders' devastation of the THFG, are marked with red lines. From top to bottom, the coordinate represents the relative positions of these gardens in the THFG region.

Blocks of different colors stand for the 8 main imperial gardens in the THFG region. The broadness of the blocks stands for their land coverage. The length stands for the time over which they existed. The gray color stands for garden ruins.

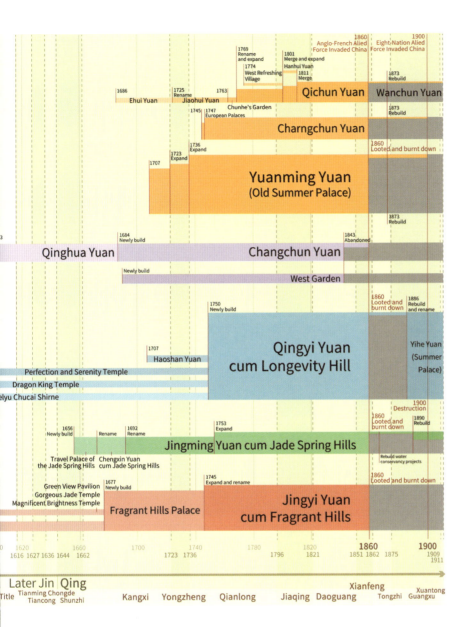

# IV Illustration of history: The millennial time axis of the THFG

## Examples

1. If you want to know about the THFG in Kangxi period (1662—1722): At that time, the Jade Spring Hills Palace had already been built by Emperor Shunzhi. Emperor Kangxi first built the Fragrant Hills Palace in 1677, put up Changchun Yuan in 1684 and the accessory West Garden soon afterwards. He granted Ehui Yuan to Fuquan in 1686, renamed Chengxin Yuan "Jingming Yuan" in 1692, and granted Yuanming Yuan to his 4th son Yinzhen and Haoshan Yuan to his 1st son in 1707.

2. If you want to know about the THFG from 1860 to the end of the Qing dynasty: After looted and burned by the Anglo-French Allied Forces, the THFG was almost reduced to ruins (Changchun Yuan had already been gradually torn down since 1843). In 1873, Emperor Tongzhi ordered to reconstruct Yuanming Yuan, but this attempt went bankrupt soon afterwards. The surviving scenic spots in the garden were destroyed in the 1900 warfare. Qingyi Yuan and Jingming Yuan were restored in Guangxu period. The former was renamed Yihe Yuan (Summer Palace).

# Touring the Three Hills and Five Gardens

Landscape, Art and Life of
Chinese Imperial Gardens

# Chapter 1
# Overall Characteristic of the Imperial Gardens Region

# 1.1 The Water Yards

**In this section, you will get an idea about:**
1. When did the earliest landscape development in the THFG start?
2. In what natural environment and cultural background were the Qing imperial gardens in the western suburbs constructed?

What is landscape architecture? For most people today, it means parks or gardens, something that makes our environment more beautiful or our life more enjoyable, without or with little connection to spiritual pursuit. In reality, however, it has been attracting literatus over the thousands of years following the emergence of Shang and Zhou periods. Not only does it represent a comfortable living environment, more importantly, its scenes are impregnated with culture and they embody the spiritual appeal of the ancients. The construction of the Qing imperial gardens, notably the "Five Gardens", spanned more than 200 years from the beginning up to the end of the Qing dynasty (1644—1911). They are not only the final masterpieces of the ancient Chinese feudal dynasticism, but also a comprehensive expression of mature artistic skills in the history of landscape architecture. Their long cycle of creation and great variety of landscaping contents will undoubtedly drop us clues on the true mindsets of the generations of monarchs as well as the art of gardening that has been carried forward and refreshed since the ancient times.

## How the THFG took its name?

"TH", namely, three hills, refers to (from west to east) the Fragrant Hills, Jade Spring Hills, and Longevity Hill. "FG", namely five gardens, refers to (in chronological order) Changchun Yuan (1684), Jingming Yuan (1692), Yuanming Yuan (1709), Jingyi Yuan (1746), and Qingyi Yuan (1750). The inter-relation between the TH and the FG is the hills and the gardens seated on them: Jingyi Yuan is seated on the Fragrant Hills, Jingming Yuan on the Jade Spring Hills, and Qingyi Yuan on the Longevity Hill. The name THFG does not originate from official nomenclature, but is a vulgo used by the civilians in the late Qing dynasty to refer to the group of imperial gardens in the western suburbs. At first, it was called "FGTH", since the gardens were more politically important. Later it became "THFG", as this seems to run more smoothly. Today, this term has already become the official definition of this area. In reality, as the three travel palaces at the Three Hills—Jingyi Yuan, Jingming Yuan and Qingyi

Yuan—were equally important, the imperial family would often use the names of the hills for the gardens on them and vice versa. For example, reference to the "Three Hills" could also mean the "Three Gardens".

The garden history in the THFG region, although reaching its zenith in the Qing dynasty, dates back to as early as the Fragrant Hills and Jade Spring Hills in the Jin dynasty (1115—1234). To trace the development of the entire area, we need two important clues: the emergence of religious culture at the Western Hills during Liao and Jin periods and the boom of farming culture in Haidian during Yuan (1271—1368) and Ming periods (1368—1644). Both peaked at the end of the Ming dynasty, making the northwestern suburbs of the capital city a place of interest for all walks of life. This laid good material and cultural basis for the prosperity of imperial gardens in the Qing dynasty.

## Religious culture and the Western Hills water yards

Ancient Chinese believed in natural landscapes and benefited from them both physically and spiritually. The oldest hills in the THFG, the Fragrant Hills and Jade Spring Hills, along with the two parkland-like "upland gardens" built on them, namely Jingyi Yuan and Jingming Yuan, are the most typical examples of this philosophy. The former is renowned for its spectacular autumn views of red-hued tree leaves; the latter was well admired by the ancients for its spring water that is **"light mass with sweet taste"**. Why, you may ask, did they win favor even ten centuries ago? Well, perhaps we'll have to find the answer from history.

In Northern Song period, two minority-founded powers, Liao (907—1218) and Jin (1115—1234), coexisted in north China. The place where Beijing is now located used to be one of the five Liao capitals, the south capital called Xijin Prefecture. As one of the auxiliary capitals to the upper capital Linhuang (presently Bairin Left Banner of Inner Mongolia), Xijin Prefecture was the most prosperous region in Liao's reign. As early back as during this time, some bigwigs took a shine to this "geomantic treasure place" of the Fragrant Hills. Imperial Minister Alemi, for example, once set up a private estate in what is now the Temple of Azure Clouds. Later on, after the Jin invaders took over the south capital, they built up a capital of the greatest importance among the "Five Jing Capitals"—the middle capital Daxing Prefecture—in 1153 A.D., marking the beginning of Beijing serving as a capital city. Thirty-three years later, in the 26th year of Dading Period (1186), the Fragrant Hills became home to a imperially-funded Buddhist temple built on the former Tang dynasty ruins, the Grand Temple of Eternal Peace (also known as the Amrta Temple or Fragrant Hills Temple) at the Fragrant Hills. The inviting woods and water supplies at the Fragrant Hills offered ideal monastic venues for the Buddhists. The name "Eternal Peace" also conveys the monarch's wish to stabilize their power by virtue of religious beliefs. This represents the earliest construction of an imperial temple.

During the reign of Jin emperor Wanyan Jing, the economy and culture of the country both flourished. This period is known as "Rule of Mingchang". Meanwhile, this Nyuzhen emperor had high literacy of the Han culture and was fond of touring and hunting. Although he held in possession a grand capital town and many palaces inside it, he was

❶ A travel palace is a place where the emperor stays on short trips. Some people suggest that the other 6 water yards were the Clear Water Yard at the Temple of Great Awakening, the Fragrant Water Yard at the Temple of Cloud-like Dharma, the Devine Water Yard at the Seclusion Temple, the Sacred Water Yard at the Huangpu Temple, the Dual Water Yard at the Temple of Dual Springs, and the Golden Water Yard at the Temple of Golden Hill.

❷ A great artist and calligrapher of the Ming dynasty who used to serve at Beijing Imperial Academy in Jiajing period.

still not contented. In around 1190, he selected 8 graceful temples at the Western Hills to build his travels palaces❶. These places are now known as the "Eight Water Yards of the Western Hills". Among them are the Pond Water Yard at the Fragrant Hills and the Spring Water Yard at the Jade Spring Hills. That was the earliest construction of imperial gardens in the THFG. Historical record also says that from 1190 to 1207, Wanyan Jing toured these places for many a time. During the same period of time, the emperor also identified what are well-known today as the "Eight Scenes of Yenching". Obviously, the Jin people were already scoping the fine suburban scenes of different seasons and features into the imperial pleasure compound and giving them poetic names. For example, the Western Hills Covered by Snow and Hanging Rainbow of Jade Spring Hills scenes were located on the Fragrant Hills and Jade Spring Hills, respectively. A more detailed account of their development in the Qing dynasty is included in Chapter 2 of this book.

So, what kind of a garden is a water yard? Now that water originates from springs, a "yard" is an upland garden containing small water surfaces. For example, the Lotus Palace at the Spring Water Yard was located at the Jade Spring Hills. Emperor Wanyan Jing was said to have stayed there to escape from summer heat. The Pond Water Yard was located next to the Fragrant Hills Temple. As recorded in ancient texts, "**The terrace, pine tree and spring of Emperor Wanyan Jing is called Star Altar, Escorting Pine and Dream Spring**". The emperors built the gardens in the distant suburbs away from the capital for no other reasons. On the one hand, they needed a place to take a break in on their trips; on the other hand, sufficient, refreshing spring water is not only a must of life, but also a nutrient to the body and heart. Being also from the same clan of Nyuzhen, the Qing emperors were keen on the THFG region, possibly out of their ethnic customs and the garden-based governance tradition of the Central Plains monarchs. Each year, they would spend a long time traveling between the capital city and Haidian, and even places outside Beijing—Rehe (currently Chengde), Mount Pan (Tianjin), Mount Wutai (Shanxi), and Eastern and Western Mausoleums of Qing Dynasty (Hebei). This way, being outside Beijing or on the way became a normal routine for the emperors.

After Ming emperor Zhu Di (reign: 1403—1424) moved the capital to Beijing, the

Forbidden City and West Palace (i.e., the North Sea, Middle Sea, South Sea) once served as the main venues of events for the imperial family. The Ming Tombs in Changping was another place the emperors would go to worship their ancestors. The Golden Hill north of the Jade Spring Hills is home to another group of imperial tombs. As documented in historical texts, "**Those princes and princesses that passed away young are jointly buried in Jinshankou. It is comparable to the Emperor Jingtai's Mausoleum. The concubines are mostly buried here, too**". Its position is obviously inferior to the Ming Tombs. As depicted by *A Panorama of Emperor's Outing and Return of Worshiping Ancestor* (Fig.1.1), a long scroll of pictures depicting the ancestral worship procession of the Ming Emperor Wanli (reign: 1563—1620), a formidable array sets out on horseback and returns on boats. On their way back, they stop somewhere around the West Lake (former name of the Kunming Lake) at the Jade Spring Hills and Urn Hill (former name of the Longevity Hill) to take a boat cruise. Then they cruise the Long River to the Xizhi Gate. The picture portraits a bustling suburban scene. Gardens, temples, mausoleums, and farmlands are densely distributed at the foot of the hills; the evergreen trees in the mausoleums, willows along the river banks, and peach blossoms in the gardens refresh these artificial elements with vigor. The depiction coincides with what has been described in literature. The Temple of Gorgeous Jade, Green View Pavilion and Temple of Grand & Brightness at the Fragrant Hills, the Avatamsa Temple at the Jade Spring Hills, and the Emperor Jingtai's Mausoleum of Ming Dynasty and Temple of Merits & Virtues (first built in the Yuan dynasty) nearby came into existence during this time. Their beautiful sights and endless pilgrims attracted numerous visitors there. As recorded in the *An Overview of Sceneries in the Imperial Capital*, "**Of all the sights throughout the capital, the Fragrant Hills Temple should be No. 1**". More than that, even literati like Wen Zhengming[2] (1470—1559) were enchanted by the beautiful sceneries. They composed many verses and paintings in praise of them. Although in the 29th year of Jiajing's reign (1550), this place was violently damaged by the Mongolian Altan Khan's troops, it never diminishes the importance of the gorgeous sights of the western suburbs in the mind of the people at that time.

**Fig.1.1**   Ming, Court Painter. Part of the *A Panorama of Emperor's Outing and Return of Worshiping Ancestor* (Palace Museum in Taipei)

## Watery kingdoms and private estates

Today's Haidian District is a modernized metropolis. Few people know that in ancient times, it was also famous for the superior natural conditions and developed agricultural production. The name "Haidian" used to refer to vast expanses of lakes and wetlands. Later on it evolved into the name of this place. This old, vast stretch of water was called "Danling Pan". Emperor Kangxi once marveled at how immense it was and how splendidly it blended with the Western Hills. Obviously, it was already the most distinctive natural sight of Beijing, as well as an indispensable guarantee for agricultural production. As revealed by geographical studies, this was the water surface remnant from the previous watercourse of the Yongding River. Its wetland landscape was already in place 2000 to 3000 years ago. In and around Wanquan Village, the water surface has survived thanks to the recharge from artesian flows and natural precipitations.

In North China where the land is mostly arid, people are good at farming; in Jiangsu and Zhejiang where water networks are densely distributed, people are proficient in reclaiming land from lakes. From Yuan dynasty to Ming dynasty, the paddy-dominated agricultural production in and around Haidian boomed and even attracted farmers from the regions south of the Yangtze River to find new opportunities. When the socioeconomic development has reached a substantial level, the suitable ecological environment will usually give a push to the emergence of gardens (Fig.1.2). In Wanli period of the Ming dynasty (1572—1620), the emperor's maternal and powerful grandfather Li Wei, and prestigious calligrapher and artist Mi Wanzhong each built a well-known garden northwest of Haidian Town. The former was named Qinghua Yuan (Garden of Clear Water and Flourishing Woods), the latter Shao Yuan. These two gardens were reputed as the **"most prosperous of the garden enclosures in the capital city"**. The *An Overview of Sceneries in the Imperial Capital* appraises them as: **"Li's garden is grand, Mi's garden circuitous. Neither bears the smell of feudal scholar's malady"**. Both gardens were continued and developed in the Qing dynasty. Qinghua Yuan was to become Changchun Yuan, as will be described in Chapter 3; Shao Yuan was to become an office garden—Hongya Yuan (Garden of Justness and Elegance) (also called Jixian Yuan, or Yard of Gathering the Virtuous).

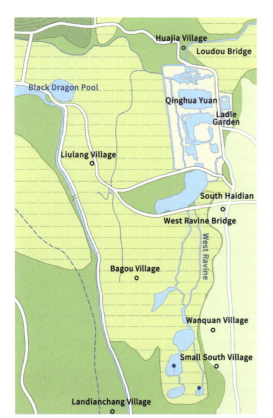

Looking back, if we say that the "Eight Water Yards of the Western Hills" were the pioneers of gardens in the THFG, the private gardens present in the Ming dynasty should have been striding toward maturity. In regions south of the Yangtze River, especially in Suzhou and Wuxi, private gardens were already maturing. There famous gardens like Zhuozheng Yuan (also known as the Humble Administrator's Garden), Yipu (also known as the Garden of Cultivation) and Jichang Yuan (also known as the

Garden of Placing Ease) came into being right at this time.

Shao Yuan and Qinghua Yuan distinguished themselves from the Jin dynasty Water Yards not by size, but by site selection and landscaping approach. The area around Haidian characterized mild terrain and rich water. If a suitable piece of land was marked out and manually reworked, it should be highly malleable. Yet it was also extremely challenging to the ability of the landscape architect to create scenes with natural resources. By comparison, if natural hills were used to build gardens, the malleability wouldn't be so high. After all, the ancients' ability to modify landform was very limited. They would have no other way than make local modifications to accommodate house building. Besides, it was also impractical for an upland garden to provide large lake surfaces or rich sightseeing experience. Both Mi Wanzhong and Li Wei, one a descendant of an artist family and the other a financially sound celebrity, chose to build their own gardens in the plain wetland next to Haidian Town. Although Qinghua Yuan had been heavily criticized for being excessively sumptuous, at large, both gardens possessed a very high artistic taste. This was attributable to the fusion of gardening arts between the north and the south. The smart craftsmen from regions south of the Yangtze River brought in with them their mature artistic approach of hills and waters placement. For example, the book *The Craft of Garden* authored by Ming dynasty "philosophical garden builder" Ji Cheng (1582—?) is a textbook of the greatest influence.

## Qinghua Yuan and Shao Yuan

The name Shao Yuan (also known as the Ladle Garden) signifies that the garden was as small as a ladle compared to the vast water surfaces in and around Haidian. Its actual size was about "100 mu" (c. 66,000 m²). This garden used to be a good place for literati to get together. In the 45th year of Wanli's reign (1617), Mi Wanzhong painted *A Panorama of Waterside Ritual in Shao Yuan* to record the garden landscape at that time(Fig.1.3). The garden had quite a large area of waters, but it looked extremely deep with the causeways, islands and bridges. On top of these structures were placed buildings of varying sizes and shapes. The overall sight largely resembled a garden typical of the regions south of the Yangtze River.

In contrast, immediately across from the road, Qinghua Yuan was often said to be "**extremely grand**". Not only did it boast an overwhelming advantage of being 10 li (5000 m perimeter) around, its tens of li of waterway tour line also offered a magnificent sight of lakes, tall buildings, rockeries, plants and animals. The extreme extravagance once triggered a lot of complaints in the verses of many people at that time.

**Fig.1.2**
Distribution map showing the water systems of the gardens in Haidian in the Ming dynasty

**Fig.1.3**
Ming. Wu Bin. Part of the *A Panorama of Waterside Ritual at the Ladle Garden* (Peking University Library)

# 1.2 Scale

From the restored plan, one can easily see that the 5 large imperial gardens intended to be directly toured by the imperial family in the THFG region, namely, Yuanming Yuan, Changchun Yuan, Qingyi Yuan, Jingming Yuan, and Jingyi Yuan, did not exist in the northwestern suburbs of Beijing by themselves. Accompanying these imperial gardens were many government offices for local administration, smaller granted gardens for the emperor's family members or ministers, villages and paddy fields for civilians to reside and toil on, barracks for the safety of the imperial gardens, and camps of special forces. Obviously, in the Qing dynasty, the population structure in the THFG region was quite complicated. There were not only the imperial family, ministers and armymen, but also monks, servants, craftsmen, farmers, and tradesmen. In fact, it encompassed people of all occupations from all walks of life. Also, as it was directly administered by the imperial family and constantly resided by the imperial family members, it well deserved the name of "Lesser Capital".

Although it neither was it clearly bounded by a city wall and a moat around it, nor was integrally controlled by a 7.8-kilometer long central axis like Beijing, the THFG region was indeed a suburban "city" without a city wall. Regarding its exact scope and area, although undetermined yet, if we tried to link the circularly distributed Guard Division of the Eight Banners Forces of Yuanming Yuan together, we would find that all the hills and gardens, with the exception of only the Fragrant Hills cum Jingyi Yuan, were seated in a circle with a 8,000 m diameter (Fig.1.4). The center of this circle fell in the Imperial Racecourse west of Changchun Yuan. If we wanted to contain the Special Division of the Eight Banners Forces and all temples around Jingyi Yuan, a circle with a 5,000 m diameter would be needed. If that should be the case, the two circular locations should actually include almost all the contents of the THFG region, including the 8 elements of hills, woods, waters, gardens, fields, villages, temples and barracks.

If we are to calculate its overall area, the following formula may be used: Area $(S) = \pi \times R^2$.

The total area of the THFG region should be no smaller than $50.265 + 19.635 \approx 69.9$ km² > 62.5 km² (the area of the inner and outer Beijing city).

In all, there were 10 imperial gardens and 14 granted gardens (Tab.1.1). The total area was 1175.6 hm². The total area of these gardens was about 16 times the size of the Forbidden City.

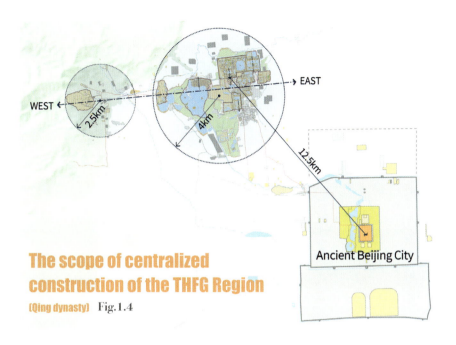

**The scope of centralized construction of the THFG Region**
(Qing dynasty) Fig.1.4

Tab.1.1 Sizes of the gardens in the THFG Region (Xianfeng period)

| Type | Name | Scale (hm²) |
|---|---|---|
| Imperial Garden | Changchun Yuan | 54.6 |
| | West Garden | 31.5 |
| | Yuanming Yuan (Old Summer Palace) | 207 |
| | Charngchun Yuan | 75.6 |
| | Qichun Yuan | 70.6 |
| | Qingyi Yuan | 295 |
| | Jingming Yuan | 77 |
| | Jingyi Yuan | 156.5 |
| | Quanzong Miao | 2.1 |
| | Temple of Imperial Moralization | 37.2 |
| | **Total** | **1007.1** |
| Granted Garden | Jinchun Yuan | 24.1 |
| | Tsinghua Yuan | 28.1 |
| | Minghe Yuan | 8.2 |
| | Langrun Yuan | 7.6 |
| | Jingchun Yuan | 1.7 |
| | Weixiu Yuan | 8.2 |
| | Chengze Yuan | 2.3 |
| | Shuchun Yuan | 25.5 |
| | Chenghuai Yuan | 3.8 |
| | Prince Li's Garden | 2.4 |
| | Prince(Beizi) Zhi's Garden | 14.1 |
| | Jixian Yuan | 4.5 |
| | Zide Yuan | 14.7 |
| | Chunxi Yuan | 23.3 |
| | **Total** | **168.5** |

**Note**
Changchun Yuan and West Garden were already out of use in Xianfeng period

# 1.3 Layout Features

**In this section, you will get an idea about:**
1. What are the distinctive geographical features of the THFG region?
2. Which water systems were once the "blood veins" of the THFG region?
3. What is the scientific principle behind the siting and layout of the imperial gardens?

So, in front of such a sizable area, how can we interpret the characteristics of its layout? In general, the following two observations best describe these characteristics. First, an east-to-west transition from being urban toward being natural; second, an interconnection of eastward, westward, and southward water flows.

## Feature 1: Three different landscapes from east to west

In the plain area in the east, Haidian Town, Changchun Yuan, and Yuanming Yuan were surrounded by a group of granted gardens—Shuchun Yuan, Weixiu Yuan, Minghe Yuan, Chenghuai Yuan; government offices—the Imperial Opera Department and Imperial Wood Factory; hamlets—Chengfu Village, Liulang Village, Guajia Village; and temples—the Temple of Listening to Buddha Dharma, Temple of Benevolence and Blessing, and Nunnery of Fine Affinity. These establishments were separated from each other by streets, bounding walls or waterways in good order, making up an enormous "garden community".

We can try to picture setting out from the capital city's Xizhi Gate and walking about 6 km along the imperial road to reach the populous Haidian Town. Here, everything is bustling with noises and excitements. Stores stand in great numbers. Quite a few of them are the time-honored brands from the capital city. The place is also densely covered with private estates and gardens of celebrities. This precisely answers to the account of "**red doors and green tiles, rows upon rows of houses, with no difference from inside the city**".

Walking along the South Street through the end and turning to the West Street, we arrive at Changchun Yuan. If we walk on the imperial road east of this garden, we will see that both sides of the road are almost all covered with bounding walls of the gardens. These gardens are properties of either the imperial family members, like princes or princesses, or favorite ministers of the emperors. As they are not hereditary❸, these granted

❸ Historical texts show that the granted gardens were also planned and constructed by imperial artisans. They were actually the property of the emperors.

gardens have been transferred from one owner to another. Further north, after crossing the Grand Rainbow Bridge, we find ourselves in the territory of Yuanming Yuan. Here, the huge Fan Lake and spacious Grand Palace Gate plaza combine to become the "national plaza" of the Qing empire (see more details in the Court at Front, Life at Rear section of Chapter 5). Although this used to represent the image of the country, it does not mean prohibiting civilians from residing on its periphery. Instead, the inhabitants of the hamlets like Guajia Village and Chengfu Village were the most needed labor power for operating the imperial gardens. The large tracts of paddy fields surrounding the imperial gardens were also worked by them .

West of Changchun Yuan and Haidian Town and east of the East Causeway of the Kunming Lake, the terrain is low with plentiful spring holes. It is the best place for growing Rice of Western Beijing. Here, the hundreds of hectares of paddy fields stretched along the Wanquan River (Tens of Thousands of Springs River) and its tributaries, filling up almost the entire area. Only a small fraction of the land belonged to Liulang Village and Bagou Village and the two imperial temples: Quanzong Miao (Temple of Springs' Origin) and Temple of Imperial Moralization (Shenghua Si). The sight was extremely spectacular.

Further west was the magnificent mix of lakes and hills: the Jade Spring Hills cum Jingming Yuan and the Longevity Hill cum Qingyi Yuan. The bounding wall of Qingyi Yuan did not scope the Kunming Lake into its compass. The overall idyllic scenery was retained to the furthest extent possible. Which successfully blended the garden landscapes with the external environment (Fig.1.5-a). Here was not only another place where Rice of Western Beijing was centrally cultivated. But also an important water conservancy hub, too. The Northern Dock Village and Middle Dock Village were sporadically distributed

**Fig.1.5**

Old photographs of the 4 representative scenes (a. Jade Spring Hills with the Picturesque Scenery Archway, b.Azure Dragon Bridge, c. Western Hills and Temple of Azure Clouds, d. Long River with imperial boat)

close to the Rice of Western Beijing production area. A particular place was the Town of Azure Dragon Bridge northwest of the Longevity Hill. Remote as it might appear, it lay on the way that the emperor had to pass when traveling to the Jade Spring Hills and Fragrant Hills. More importantly, Azure Dragon Bridge had been an important facility safeguarding the water conservancy security since the Yuan dynasty (Fig.1.5-b).

West from the Jade Spring Hills, the landscape switched again to vast plains and the lofty Western Hills, taking on the typical features of north China. On the plains there were no lake surfaces but sporadic tombs and woods. Man-made pieces were mostly situated halfway or at the foot of the Small Western Hills. The Fragrant Hills cum Jingyi Yuan held the entire 130–575 m altitude valley into its embrace. Inside and on the periphery of the garden, there were also a range of imperial temples (Fig.1.5-c). Besides, Han villages and the banner barracks of the Special Division of the Eight Banners Forces stretched in patches on the left and right flanks of Jingyi Yuan. The Round City and Hall of Reviewing Troops and the drill ground were also seated nearby. More notably, dozens of Special Division of the Eight Banners Forces barbicans also stood halfway of the hill (see more details in the Military Drill section of Chapter 5). Compared with the tall hills, however, these man-made constructions were indeed a drop in the bucket. The aqueduct traversing the plains was even more striking to the eye. From east to west, subject to the natural conditions, the ancients created three different features of landscape: city, farmlands and hill forests.

### Feature 2: Three relatively independent water systems

The three water veins in the THFG region are (Fig.1.5): the NS-trending Wanquan River water system, the EW-trending Western Hills–Jade Spring Hills–Longevity Hill

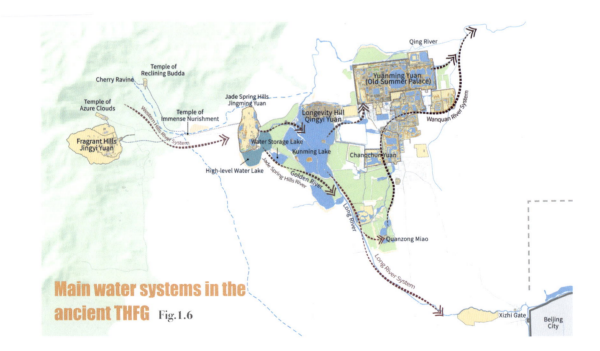

**Main water systems in the ancient THFG** Fig.1.6

water system, and the NW–SE-trending Long River water system (Fig.1.5-d). Due to geological reasons, numerous spring holes are present at the Western Hills and Jade Spring Hills in the THFG and at Bagou Village in the plains. Each of them has formed their respective water systems under the combined forces of man and nature. As we will introduce in details about the water conservancy engineering in the western suburbs in Chapter 4, here we will only give a brief introduction on the Wanquan River water system and its unique values.

The Wanquan River originates from the myriads of springs around Bagou Village and Wanquan Village. As noted in ancient texts, "**Countless springs gurgle out of the small holes**". They assembled in the Lingjiao Pond (Water Caltrop and Water Bamboo Pond, a lakelet at the location of ancient Danling Pan) south of the Main Palace Gate of Changchun Yuan, then ran through the gardens in Haidian and separated into a number of rivulets. The river channel bent eastward at the northern tip of Weixiu Yuan (Garden of Deep Luxuriance and Grace), supplying water into Qichun Yuan, Langrun Yuan (Garden of Brightness and Moisture) and Xichun Yuan (Garden of Prosperity and Spring) (Fig.1.7). In

**Fig.1.7**
Waters scenes at Wanquan River and Charngchun Yuan

the meantime, it also branched off into Chengfu Village and Haidian Town. Finally, at the northwest of Xichun Yuan, it "**trends toward the Qing River together with the streams from the imperial gardens, joins the Sand River and pours into the White River**". The entire river covered a distance of about 9 km. To ensure adequate water supply, the Qing people installed a culvert on the eastern bank of the Long River to discharge part of the water into the Quanzong Miao and redirect it northward. After the Kunming Lake water conservancy hub was established, the stored water would also supply water into the gardens and paddy fields on the east side before pouring into the Qing River and heading eastward. In the 32nd year of Qianlong's reign (1767), a tiny imperial temple garden—Quanzong Miao—appeared at the fountainhead of the springs. This garden was intended for both religious worship and pleasure. In a way, the Wanquan River basin was cradle to the THFG. The life, production, and garden-landscaping there could never have proceeded without its constant supply of water.

Geographically, such a layout stood out from others due to the following facts: One is that the ancients, with limited engineering capability, were able to harness such a semi-natural, semi-artificial water system and cause it to serve the human settlement. Second is that the site was selected rationally. Haidian Town, Wanquan Village and other hamlets were all situated on elevated grounds at least 50 m above sea level (Fig.1.8) . Their high terrain could protect them from flooding. However, the gardens were almost all seated on low-lying grounds less than 50 m above sea level. They started with Changchun Yuan in the south and terminated at the Qing River in the north, the Wanquan River in the east, and the East Causeway of the Kunming Lake in the west. The gardens were close enough to make the best use of the water availability for gardens situated in lowlying places. It's fair to say that the Wanquan River basin and the natural conditions in its proximity cast the splendor of the Qing imperial gardens. In return, the construction of these gardens and water conservancy facilities, especially drainage excavation and vegetation planting, greatly improved the ecological environment there.

On these grounds, it won't be hard to speculate how the gardens in this area evolved in the Ming and Qing dynasties. The Ming dynasty Qinghua Yuan and Shao Yuan were located on a suburban lowland very close to Haidian Town. They were two private gardens based on Danling Pan. In Kangxi's reign of the Qing dynasty, Changchun Yuan emerged to continue with the legend of Qinghua Yuan (see more details in Changchun Yuan section of Chapter 3). Led by Emperor Kangxi, a succession of granted gardens of the imperial family members and ministers sprouted in the north and east. They all circled around Changchun Yuan like stars twinkling around the moon (see more details in the Clustering of Gardens in Haidian section of Chapter 3). At that time, except for Changchun Yuan and West Garden that were large, 880,000 m² imperial gardens, all the other gardens around them were of medium or small sizes. After Emperor Yongzheng ascended the throne, he extended his mansion Yuanming Yuan all the way north to the fringe of the Qing River. He also opened a huge Blessing Sea scenic area in the east. In Qianlong–Jiaqing period, Charngchun Yuan and Qichun Yuan even occupied large tracts of empty land and village land in the Wanquan River basin. This way, Yuanming Yuan, along with 4 accessory gardens, grew into an ultra-large imperial garden complex. The land in the basin available for building gardens became very scarce. The layout of the THFG was finalized at that point, too. Little change had taken place until the outbreak of the 1860 war. Frustratingly, half a century before their destruction, these gardens were already showing noticeable signs of decay. Even the older-aged imperial garden of Changchun Yuan ended up being discarded. Its timber was torn down for building other gardens. What a shame!

Presumably, the imperial gardens, villages and farmlands in the Wanquan River basin were a meticulously structured, complete organic body that served the then political center of the empire. It actually surpassed any single imperial garden in ecological, cultural, and artistic values. The scientificity and artistry behind its general layout has yet to be fully identified. Further attention and protection is still necessary.

Looking back on history, before Changchun Yuan was completed in the early Qing dynasty, the imperial family were simply building their travel palaces at the Fragrant Hills and Jade Spring Hills on their ruins from the previous dynasties. The identity of the THFG was not formally established until Emperor Kangxi officially lodged in Changchun Yuan in the 26th year of his reign (1687). After that, the 4 generations of emperors—Kangxi, Yongzheng, Qianlong, and Jiaqing—each invested unremitting efforts launching water conservancy works, opening farmlands, building gardens, developing religions and deploying military troops. In the 16th year of Jiaqing's reign (1811), the expansion of Qichun Yuan finally declared completion, putting an end to this prolonged 127-year great cause. Up to the end of the Qing dynasty, it had operated for an aggregate of over two centuries or 85% of the length of reign of the Qing dynasty.

# 1.4 Garden-making procedure

**In this section, you will get an idea about:**
1. Why are the artificial hills and waters at Yuanming Yuan equivalent to an "engineering miracle"?
2. What are the characteristics of the buildings in Chinese imperial gardens?

After taking a careful inventory, we can discover that the entire THFG region was an organic mix of 8 natural and artificial elements: hills, woods, waters, gardens, fields, villages, temples, and barracks. The natural or semi-natural elements were hills, woods, and waters; the artificial ones were gardens, villages, temples, and barracks.

## Use of natural landscapes

In the THFG region, natural hills primarily included the Small Western Hills, Jade Spring Hills, and Longevity Hill. The ancients named a number of peaks in the Small Western Hills. Among them, the most famous ones were the Fragrant Hills (Incense-Burner Peak), Red Hill (Red Rock Hill), Longevity and Peace Hills, and Wang'er Hills (Baiwang Hills). They jointly composed the landscape framework of this area (Fig.1.8). If we look at the orientation of the hills, the crest line of the Small Western Hills stretches from south to north, then bends to the east and terminates at the Baiwang Hills as its eastern tip. On this main vein, the hill also stretches sidewise into a number of branches. Among these branches, the most distinctive ones should be the Red Hill and Longevity Hill. Although they are 500 m apart, the two hills are invisibly connected beneath. In the same way, the Jade Spring Hills is also one of the branches of the Small Western Hills, except that this hill is more undulated and contains many springs both up and down. For this reason, the Jade Spring Hills and Longevity Hill are called "relict hills" (Fig.1.8). Accordingly, despite the different geological locations, the "Three Hills" actually belong to the same one hill (Fig.1.9).

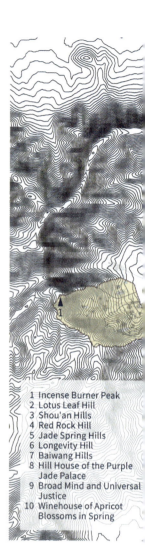

1 Incense Burner Peak
2 Lotus Leaf Hill
3 Shou'an Hills
4 Red Rock Hill
5 Jade Spring Hills
6 Longevity Hill
7 Baiwang Hills
8 Hill House of the Purple Jade Palace
9 Broad Mind and Universal Justice
10 Winehouse of Apricot Blossoms in Spring

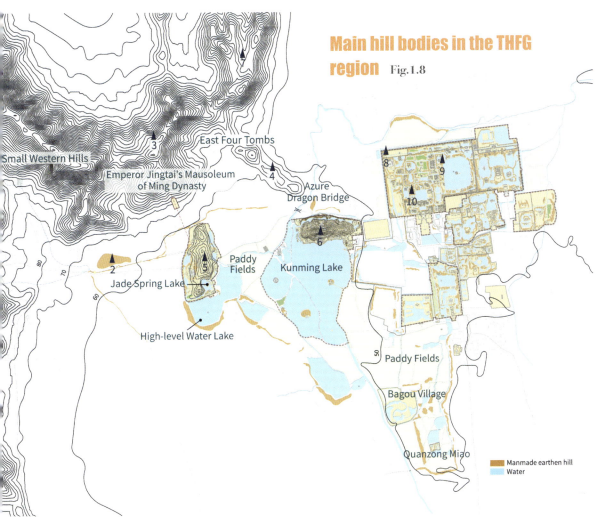

**Main hill bodies in the THFG region** Fig.1.8

Fig.1.9
An overlook at the Jade Spring Hills and Western Hills from the Longevity Hill (1902)

Morphologically, the Small Western Hills look like two arms reaching out southeastward to "embrace" the entire THFG region. Hence the towering landform also functions as the best natural hedge against northwest wind. It played a critical role in the formation of the unique water-nourished wonderland in this part of the suburbs. In other words, the hills and springs are both indispensable. The Ming imperial family, after examining the landscape situations here, decided that it accorded well with their requirements for building their tombs. As can be observed from the remains of the Emperor Jingtai's Mausoleum of Ming Dynasty, the axis line of the palace points straight to the Jade Spring Hills. Besides, as the Azure Dragon Bridge lies to the southeast of the hill range, its name is possibly related to the traditional geomantic conception of "azure dragon to the left, white tiger to the right".

The natural hill bodies of the "Three Hills"—the Longevity Hill, Jade Spring Hills and Fragrant Hills—have also been artificially modified during garden-making process. When building the Kunming Lake, in order to maintain earthwork equilibrium, Emperor Qianlong deliberately used the mud excavated to dress up the Longevity Hill. Besides, he also ordered to dig an east-through-west Rear-Stream River and pile up the mud on the northern bank as earth mounds. This way, the Longevity Hill was made into a huge "island hill". The modified Longevity Hill is not only a south-facing, regularly-shaped, water-surrounded massif. Its slope gradient is also richly variable. The superb topographical design gives divergent landscaping atmospheres and experiences between the front and back of the hill, creating the foundation for the temples and gardens up and down.

As to the Fragrant Hills and Jade Spring Hills, in order to utilize the spring water in the hills and create landscapes, only a few local terrain modifications were done to the hill bodies. For example, the terraces where the Fragrant Hills Temple is seated were manually laid with earth from cutting the hill body. At the Modesty and Openness Studio, artificial earth mounds were put up with mud from lake excavation. Comparatively, the water bodies under the Jade Spring Hills were vastly modified. Spring holes at different locations were expanded into gracefully shaped lake surfaces connected by watercourses. The riverbanks were adorned with paddy fields---that was the Supervising Plowing by the Stream, one of the "Sixteen Scenes of Jingming Yuan". The Jade Spring Lake in the south of the garden (where the First Spring Under Heaven is located) was remodeled as a key scenic area. There were scenes for viewers both on the 3 islands in the lake and around the periphery of the banks. The foothill on the western bank was almost all manually modified. Palaces were distributed at different altitudes, too, compounding a rich stratification of landscape.

Another major terrain modification was the vast stretch of paddy fields. The water conservancy works in the western suburbs contributed enormously to an agricultural boom. The colossal cascade water storage system greatly expanded the original natural lake surface (see more details in Chapter 4). It offered abundant water resources for the farmlands of paddy rice, pampas grass, lotus root and rape flower. As the paddy fields were low-lying, piles of manually excavated mud were gathered together and stacked into long strips of inflexible earth mounds to encircle the paddy fields. Some of the examples can be seen around Bagou Village and Quanzong Miao. Although structured in the same way as those inside the imperial gardens, as the earth mounds outside the gardens were built to earmark the land property of the imperial family, they looked more like a wall. During the late Qing and ROC periods, as the governments were inefficient in managing the water systems and paddy fields in the THFG, the

High-level Water Lake and the large lake surfaces inside the imperial gardens were encroached by paddies. Although the rapid sprawl of Rice of Western Beijing may have suggested of a booming agriculture on the surface, this caused severe disruption to the High-level Water Lake as a water storage facility. As the paddy fields consumed a large stock of water resources, this change once compromised the water supply to the capital city downstream and also led to changes in the sight of the landscapes here.

## "Fever" about artificial hills and waters

Compared with the modifications on natural hills and waters, what amazes us more was the creation of hills and waters in the east of the THFG region. Originally, the area in and around Haidian used to be a wetland habitat with lush trees and broad water surfaces. Houses built there would be a private suburban estate with elegant scenery. The problem was: there were only waters but no hills. The morphology of the waters was unorganized and invariable, too. The sight seemed to be a little tedious and was far below the literati's expectations about hills and waters. In order to mark off different scenic spots and vest the hill bodies with features of artistic taste, the sharp-minded ancient artisans managed to generate a contrast between depth and openness of the water surfaces in the gardens. They tried to create artificial lakes and hills by simulating nature. The trending of the earth mounds formed a blend between mildness and steepness. Local embellishment with rockeries also added to the loftiness of the mounds, making them look like real hills. This "real-out-of-false" approach gave rise to "city forests" on flat land. As Emperor Yongzheng said in his *Note of Yuanming Yuan*, "**It takes the interest of nature and saves the trouble of labor**". In the Qing dynasty, such an excellent garden-making tradition was exploited to its full potential.

All the hill bodies in these gardens on flat land were locally stacked with mud excavated from the watercourses. They can be rated as an engineering miracle. The hill bodies were mostly on the order of 3-5 m tall and a couple of them were taller than 10 m. But when people were in there, they would not feel the hills were inflexible or unnatural. Instead, they would feel like being deep into the hill forests. If we look at the overall setup on the plan layout, we will discover that the connection between the hills and waters was extremely intimate. They were widely variable with no duplication. Some of the hill bodies held the water surfaces tightly into their embrace (Spring Peach Blossoms of Wuling, Mr. Lianxi's Land of Happiness); some carried a narrow stream between its valleys (Great Mercy and Eternal Blessing, Vairocana's Magnificent Residence); others simply stood as an earth mound in the middle of a large lake surface (Academy of Gathering Talents) (Fig.1.10).

Regarding the treatment of water features, the ancients spared no efforts, either. As the height variation here is not as large as in the mountains, the water surfaces in these gardens are mostly still water. To compensate the lack of "**the elegance of woods and springs in high mountains**", during garden-making process, particular attention was paid to create cascaded waterfalls at different altitudes. Water sluices were also used to keep control of the water level and water direction in the gardens, such as those in Yuanming Yuan and Tsinghua Yuan. In the 18th century, the emperor order the use of the hydraulic machines from Europe to create dynamic waterscape, such as those at Graceful Scenery of the Western Peaks, Clear Water and Flourishing Woods, and European Palaces at

## Distribution of hills and waters at Yuanming Yuan and typical cases  Fig.1.10

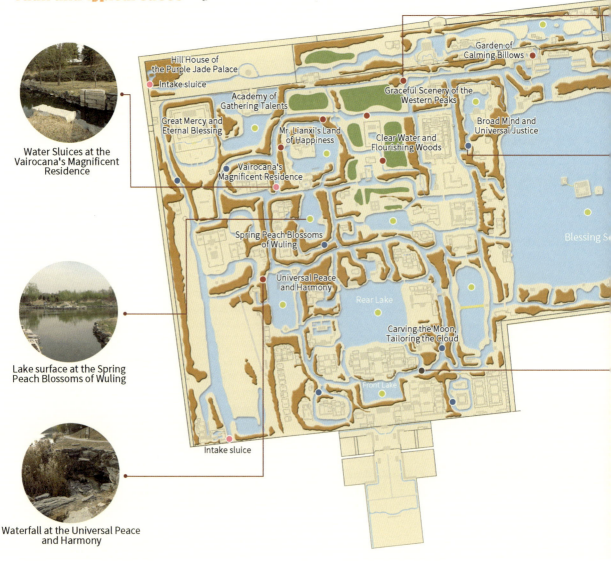

Water Sluices at the Vairocana's Magnificent Residence

Lake surface at the Spring Peach Blossoms of Wuling

Waterfall at the Universal Peace and Harmony

## Hills and waters arrangement among the granted gardens  Fig.1.11

Nine Prefectures Scenic Area at Yuanming Yuan

Chengze Yuan

Weixiu Yuan

Minghe Yuan

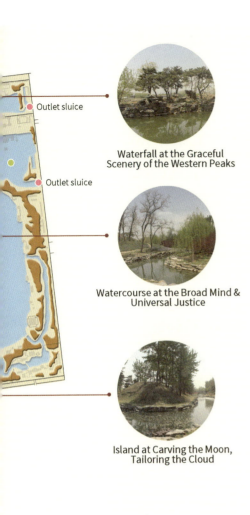

Waterfall at the Graceful Scenery of the Western Peaks

Watercourse at the Broad Mind & Universal Justice

Island at Carving the Moon, Tailoring the Cloud

Legend
- Lake
- Stream
- Waterfall
- Island
- Sluice
- Water
- Manmade earthen Hill

Charngchun Yuan. With hills and waters in place as the primary supporting structures of the gardens, building complexes of varying sizes and functions rose one after another. In other words, only in such an environment can buildings distinguish themselves by virtue of height, distance and exposure; only in such an environment can the resulted sights bear the artistic ideorealm of the Chinese landscape painting.

Regarding landscaping diversity, as the imperial gardens were overwhelmingly privileged in terms of land size and financial affordability, their landscaping and building layouts were more flexible. Their themes were more diverse and inclusive of the merits of other established gardens. For example, in Charngchun Yuan, there were a variety of themed scenes imitating the places of interest in regions south of the Yangtze River (Special Topic VI). Compared with their imperial counterparts, the granted gardens were much inferior. There, it was often hardly possible to generate such a variety of themed scenes as the "Forty Scenes of Yuanming Yuan". They were usually an assembly of a number of piecemeal scenes. From the restored plans, despite the much smaller numbers of buildings and combinations in the granted gardens, the landscaping approaches were not essentially different. Some even showed signs of stereotyping (Fig.1.11). For example, in the main lake of Weixiu Yuan, the two islands in tandem alignment were covered with the main building complexes and surrounded by earth mounds along the banks. The same was found in Jinchun Yuan (Garden of Approaching Spring). Imaginably, this dwelling environment was quite comfortable; it felt liking living on a fairly island. The Minghe Yuan (Garden of Crane Whooping), Langrun Yuan and Jingchun Yuan (Garden of Mirroring Spring) were most probably one same garden. But after separation, each of them used the main area at the center as a large island, encircled by earth mounds around them. And they resembled the "garden in garden" at Yuanming Yuan. An unusual example was Chengze Yuan (Garden of Bearing Graciousness). By virtue of the geographical advantage of the Wanquan River, the garden placed its entrance on the long causeway in the middle of the river. When visitors entered the garden, they felt like walking on waves. Obviously, the garden-making of the Qing imperial family followed a certain form of landscaping scale. Such a way of creating artificial hills and waters on flat land were probably even more challenging than modifying real hills and waters. Fortunately, in the Yuanming Yuan ruins and the Peking University and Tsinghua University campuses, these gardens still preserve astonishing numbers of hills and water scenes for people today to view and experience.

## Diverse buildings and creatures

Hills and waters constitute the basement of a garden. Building houses, growing plants and raising animals equip the garden with functions, ecological wellness and richer scenes.

Compared with their civilian counterparts, the imperial buildings were large in size, demure in shape, intricate in structure and rich in color. That is because there were many rigid official rules (Fig.1.12-Fig.1.13). Hence, not many innovations were contained in the individual houses themselves. Of course, there were still a couple of irregularly shaped buildings in the gardens, such as 卍, 田, 工, fan-shaped or boat-shaped houses. As such, what best describes the architectural character of these buildings is the diversity of combinations. Typically, corridors were used to link houses of different heights, positions and shapes, thereby giving an aesthetic impression of oneness. Quite a number of the courtyards were variants from traditional quadrangle dwellings, showcasing the creativity and imagination of the ancient artisans. In contrast, in the granted gardens, as their owners were of lower feudal hierarchical orders, the shaping of individual buildings and the form of combinations were both less flexible.

At any rate, in designing a building, the first thing to consider is its desired functions, including the feudal etiquette and religious rules. As most of the scenes were built purely for pleasure, the restrictions were still quite modest. The second thing to consider is the agreement with the theme and ideorealm of the entire scenic area. If village-themed, the buildings should be as simple and rustic as real countryside. If fairyland-themed, they should be tall and magnificent. The grouping of buildings should also incorporate hills and waters to space them apart. Attention should also be paid to how they look as a whole, including when viewed from the top and across from each other.

A livable environment is not only dependent on the furnishing inside the building itself, its exterior environment is also of crucial significance. Smart landscaping creates micro environment and micro climate; plants and animals fill the garden with life and answer perfectly to the theme of garden-making (Special Topic V). Fortunately, their utility was not subject to much constraint related to feudal hierarchy. Only the imperial family, beyond all doubt, had the exclusive right to plant rare species. Plants commonly found in the imperial and granted gardens were mostly indigenous tree species in the north. Arbors included Chinese scholar tree, peach tree, Chinese pine, oriental arborvitae, yulan, apricot, aspen and willow; shrubs included lilac, tree peony and Peking Mock-orange; aquatic plants included lotus and reed. Besides, there were numerous bamboos, vines and paddies. In the Qing dynasty, using adventitious plants and pot plants in gardens was a very common practice. According to historical texts, some rare species were sourced from many parts of the world. Chinese parasol and Mei-tree were sourced from regions south of the Yangtze River; Chinese horse chestnut, maple and grape were sourced from the western regions or beyond the Great Wall; herbs like marigold, tuberose and sensitive plant were sourced from outside China. The imperial gardens also had full-time gardeners to take care of the plants and cultivate interior ornamental plants in greenhouses. In addition to plants, the Qing imperial family would also raise many rare birds, fish, and even beasts of prey like tigers and bears. Waterfowl, deer, cranes, and brocade carp were almost the standard configuration of a garden. In short, the shapes and seasonal changes of plants, and the postures and calls of animals could all serve as a source of pleasure. Placing oneself in such an environment would feel like touring a natural hill forest.

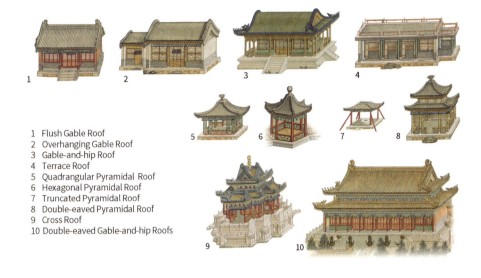

1  Flush Gable Roof
2  Overhanging Gable Roof
3  Gable-and-hip Roof
4  Terrace Roof
5  Quadrangular Pyramidal Roof
6  Hexagonal Pyramidal Roof
7  Truncated Pyramidal Roof
8  Double-eaved Pyramidal Roof
9  Cross Roof
10 Double-eaved Gable-and-hip Roofs

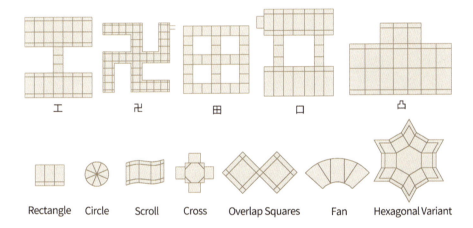

Rectangle    Circle    Scroll    Cross    Overlap Squares    Fan    Hexagonal Variant

**Fig.1.12**
Diverse roofing forms and plane shapes of ancient buildings

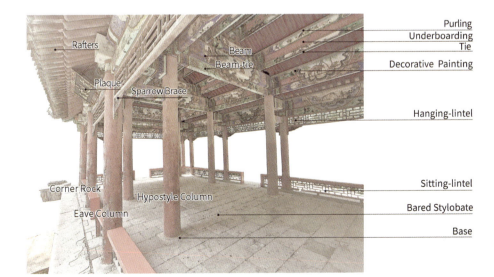

**Fig.1.13**
Basic structure of an official garden building

1

2

### [ Orientalis Arborvitae ]
*Platycladus orientalis*

A coniferous evergreen tree. City tree of Beijing. It's graceful in form, quiet in color and often planted in imperial gardens, altars and mausoleums.

4

### [ Chinese Pine ]
*Pinus tabulaeformis*

A coniferous evergreen tree. One of the most famous trees in China. It is tall, vigorous and picturesque, symbolizing constancy and longevity.

3

**"Three Friends in Cold Winter"**

### [ Mei-tree ]
*Armeniaca mume*

A broad-leaved deciduous tree. It ranks first among the famous flowers in China. It blooms in the cold of early spring, symbolizing a noble and tough character.

### [ Winter Sweet ]
*Chimonanthus praecox*

In Beijing, winter sweet is the first to open, emitting fragrance.

### [ Yellow-groove Bamboo ]
*Phyllostachys aureosulcata*

An evergreen gramineous plant. Being straight outside and hollow inside, it represents a character of integrity, fortitude, and modesty.

8

9

### [ Smoke Tree ]
*Cotinus coggygria*

A broad-leaved deciduous tree or shrub. It's a very common autumn leaf plant and named for its pink flocculent flowers.

12

### [ Broadleaved Lilac ]
*Syringa oblata*

A deciduous tree or shrub. A native plant of Beijing, which blooms in spring with purple, small and fragrant flowers.

### [ Shantung Maple ]
*Acer truncatum*

A broad-leaved deciduous tree. Its samara looks like ancient Chinese gold ingot, which explains its Chinese name "元宝枫" (Gold Ingot Maple). It is often used as a shady tree. Its leaves turn yellow or red in autumn.

13

16

### [ Peach-tree ]
*Amygdalus persica*

A broad-leaved deciduous tree native to China. Its flowers represent spring scenery. Its fruit is edible and symbolizes longevity. Its branches are said to have the magic power of suppressing disasters and warding off evil spirits.

### [ Lotus ]
*Nelumbo nucifera*

An aquatic herbaceous flower and one of China's most famous flowers. Often used as a summer ornamental in gardens. It symbolizes the noble character and also has the Buddhist implication.

17

---

### [ Sika Deer ] *Cervus nippon*

In Chinese it sounds the same as another Chinese character "禄", which means wealth. It's one of the auspicious animals for ancient Chinese with multiple connotations such as longevity and power.

1

### [ Brocade Carp ]
*Cyprinus carpio haematopterus*

A colorful ornamental fish, which symbolize auspiciousness and good luck.

4

2

### [ Roe Deer ]
*Capreolus pygargus*

A species of deer commonly found in northeast China. Some would be kept in imperial gardens.

### [ Red-crowned Crane ]
*Grus japonensis*

A symbol of longevity, auspiciousness and elegance. It is often associated with Taoist mythology.

3

# Plants

V

Common plants and animals in imperial gardens

A broad-leaved deciduous tree. City tree of Beijing. It's leafy and often used as a shade tree. The flowers are fragrant and can be used as medicine.
### [ Chinese Scholar Tree ]
*Sophora japonica*

6

### [ Weeping Willow ]
*Salix babylonica*
A broad-leaved deciduous tree. Having graceful branches, it's often planted at banks with peach. The ancients often sent willow branches to express their parting feelings.

5

### [ Maidenhair Tree ]
*Ginkgo biloba*
A broad-leaved deciduous tree. A rare relict tree species. Its leaves are fan-shaped and turn yellow in autumn. Its fruit can be used as medicine.

### [ Chinese Horse Chestnut ]
*Aesculus chinensis*
A broad-leaved deciduous tree. It has palmate seven-lobed leaves and white pagoda-shaped flowers, with a Buddhist connotation.

7

### [ Yulan ]
10
*Magnolia denudata*
A broad-leaved deciduous tree. One of the famous trees in China, with tall and straight shape and large, white and fragrant flowers.

11
### [ Chinese Flowering Crabapple ]
*Malus spectabilis*
A broad-leaved deciduous tree. An ornamental plant commonly used in courtyards. Its form is upright and the flowers are dense and delicate.

## Representative of Wealth & Auspiciousness

### [ Winter Jasmine ]
14
*Jasminum nudiflorum*
A deciduous shrub. It blooms in early spring, hence takes its Chinese name "迎春"(Welcome Spring). The branches are drooping, the flowers are small and yellow. It is often paired with rockery.

### [ Tree Peony ]
*Paeonia suffruticosa*
A deciduous shrub. One of the famous flowers in China. Being big, fragrant and rich in color, it has a laudatory name of "national beauty and heavenly fragrance" and symbolizes wealth and luck.

15

18

### [ Reed ]
*Phragmites australis*
An aquatic or wet gramineous plant. A tall, large grass often found near water. It has ornamental, economic and ecological values.

19
### [ Rape Flower ]
*Brassica napus*
A herbaceous plant. Often grown in acres, it has golden flowers and seeds that can be used for oil.

### [ Rice ]
*Oryza sativa*
A cereal crop. It has been the staple food of the Chinese since ancient times. It's also one of the main crops and ornamental plants in the Three Hills and Five Gardens.
20

6

### [ Mallard ]
*Anas platyrhynchos*
A common waterfowl with both ornamental and edible values.

# Animals

### [ Whooper Swan ]
*Cygnus cygnus*
A common kind of waterfowl, with white feathers and graceful figures. It symbolizes faithful love.

5

### [ Mandarin Duck ]
*Aix galericulata*
A common waterfowl. These ducks are seen as the symbol of love as they are always in pairs.

7

## Conveying emotions into scenes

That was how the gardens were elaborated step by step with the contents described in the previous section. They are a manually creation of a "third nature [4]"; they are a representative manifestation of "humanization of nature [5]". "Humanization of nature" means construing natural beings from human perspective and endowing them with moral characters. In the profound Chinese history and culture, the countless natural elements are given meanings beyond their essence. Weather changes, the sky, the earth, the sun and the moon, hills and waters, and plants in the natural world can all arouse man's contemplation and emotional changes. They produce a link between "scenes" and "emotions". However, these contents were less than enough for the ancients who were so passionate about expressing themselves. Hence, the ancients began to try conveying these beauties with literature and art "on their behalf". The couplets and plaques hung on the garden buildings, as well as the numerous proses, paintings, and utensils all became a vehicle of emotions to convey the ideals, pursuits and tastes of the emperors and literati.

In the well-preserved Yihe Yuan (Summer Palace), the varieties of plaques and couplets are almost a "cultural treasury". Here you will not only see "naming plaques" identifying the names of scenes. You will also see a range of "explanatory plaques" providing supplemental explanation of these names. These plaques, hand crafted from wooden or stone material, are all hung at very conspicuous parts of the buildings. The couplets on the columns, usually installed on opposite sides, constitute a rhetorical echo to the plaque. They are all important windows for interpreting a garden, a scene, a building, and even a set of rockery. This form of art makes the classic Chinese garden stand out from its peers in the other parts of the world.

[4] A generally accepted definition is that the "first nature" is the primitive wild nature; the "second nature" is modified idyllic scenery; the "third nature" is the nature manually recreated to human's subjective will.

[5] This term was first used by Karl Marx. In essence it means that nature has continuously become an existence with human attributes in practice.

### Plaques and couplets at the Lotus Fragrance Gazebo at Yihe Yuan

Naming plaque on the east:
**Ouxiangxie** (Lotus Fragrance Gazebo)
Couplets on the east:
First line (Right): **Yu Se Yao Qin Yi Tian Ban** (Zithers play to mimic sound as if it comes from the Heavens)
Second line (Left): **Jin Zhong Da Yong He Yun Men** (Bells combine to compose music for worshiping the Gods)

Explanatory plaque on the west:
**Riyuechenghui** (The bright sun and moon shine upon the world)
Couplets on the eave columns:
First line: **Tai Xie Cen Ci Jin Bi Li** (Towers and pavilions stand at different heights as if they were in a landscape painting)
Second line: **Yan Xia Shu Zhan Hua Tu Zhong** (Mists and clouds dance in various postures as if they were in a nature-made picture)
Couplets on the hypostyle columns:
First line: **Lyu Huai Lou Ge Shan Chan Xiang** (Locust trees and pavilions set off one another with cicadas chirping high and low)
Second line: **Qing Cao Chi Tang Cai Yan Fei** (Grass and water ponds add green to each other with swallows flying back and forth)

The Lotus Fragrance Gazebo is the western side hall of the Jade Billows Hall (Fig.1.14). On the east side, the plaque hung on the door header shows the name of the pavilion, "Lotus Fragrance", implying to the theme of this location as a site for appreciating lotus blossoms. This plaque is accompanied by a pair of couplets describing the elegance and magnificence of the imperial music.

On the west side of facing the Kunming Lake, an explanatory plaque is inscribed in praise of the natural beauty as well as the metaphor of peaceful and prosperous times. This plaque is accompanied by two pairs of couplets, one on the eave columns and the other on the hypostyle columns. The couplets on the eave columns are antithesis, depicting the terraces and water pavilions of the Longevity Hill and the misty water surfaces of the Kunming Lake, saying that they look like an elaborately painted picture. The couplets on the hypostyle columns at the back are quotation of Shen Kuo's verse (North Song dynasty), depicting cicadas chirping in locust trees and swallows flying over grassy ponds. They accord perfectly with the celebrative atmosphere of Yihe Yuan in Guangxu period.

**Fig.1.14**
Eastern and western facades of the Lotus Fragrance Gazebo

# VI Domestic and overseas sights hidden in the THFG

"**Shui Dao Jiang Nan Feng Jing Jia, Yi Tian Suo Di Zai Jun Huai**" (Admirable they say is the southern sight, its grace downsized to the emperor's side).

—Wang Kaiyun [6] *Poem of Yuanming Yuan*

Since the end of the Qing dynasty, this familiar line has been recited for many years. It describes how the Qing emperors had reproduced a range of places of interest in regions south of the Yangtze River up into the imperial gardens by "relocating the heaven and downsizing the earth". In fact, as projected by the four Qing emperors of Kangxi, Yongzheng, Qianlong and Jiaqing, the entire THFG was intentionally trying to reproduce the representative scenes of several provinces through the imperial landscaping approach. This approach is known as the art of "Xiefang (depict and imitatively construct)". As a matter of fact, "Xiefang" is not the invention of the Qing emperors. It is a tradition dating back to thousands of years ago. An early representative case was how the First Emperor of Qin built his own palace by imitating the palaces of the other 6 kingdoms. Civilians would also try to imitate famous mountains and great waters by creating artificial lakes and hills in their private gardens.

The THFG contained many scenes prototyped from famous scenes and gardens in regions south of the Yangtze River. They also included distinctive buildings or gardens resembling those beyond the Great Wall, in Shanxi, Shandong, Hebei, Hubei, Hunan, Sichuan, Tibet, and even Burma and Europe (where the emperors had not all visited by themselves). This cross-space and cross-time cultural phenomenon is extremely rare throughout the Chinese garden history. It offers good evidence for the openness of the imperial gardens to different forms and styles. In terms of similarity, however, considerable divergences can be detected among these scenes. Some were more peculiar about the "identity in both form and spirit"; others might simply borrow the ideorealm or name in pursuit of "spiritual similarity". As Emperor Qianlong commented in his poem, "**To imitatively construct does not necessarily mean to give up the character and style of the imperial properties themselves.**"

[6] 1833—1916, late Qing classics scholar and man of letters.

Fragrant Hills
Jingyi Yuan

Temple of Azure Clouds

Special Division of the Eight of the Fragrant Hills

Temple of Genuine Triumph
Temple of Buddha's True Essense
Temple of Buddha's Solemn Appearance
Rectangular Tibetan Barbican
Round Tibetan Barbican

India 1

East Asia

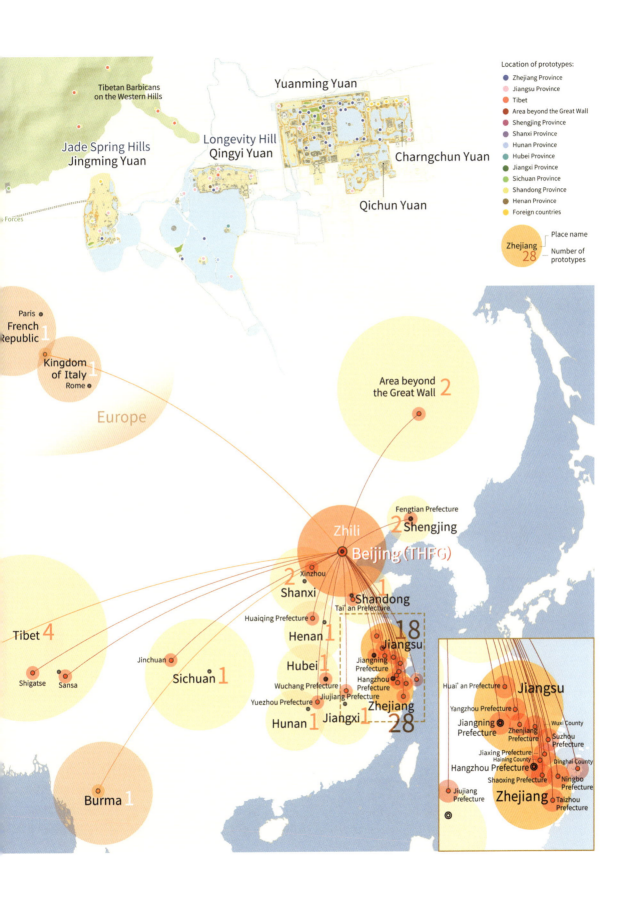

# Touring the Three Hills and Five Gardens

Landscape, Art and Life of
Chinese Imperial Gardens

# Chapter 2
# Jingyi Yuan and Jingming Yuan
—Imperial Palaces Incorporating the Charm of Woods and Springs

The exploitation of the THFG region in the Qing dynasty actually commenced from Emperor Shunzhi, the first emperor of Qing following their entry into the Central Plains. In the 13th year of Shunzhi's reign (1656), the Jade Spring Hills Palace was completed. Later on, Emperor Kangxi renamed it Chengxin Yuan (Garden of Tranquil Heart). The name means that the garden scenes were purifying and soothing.

In the 16th year of Kangxi's reign (1677), Emperor Kangxi put up a small imperial garden—the Fragrant Hills Palace (Fig.2.1)—on the empty ground next to the temples at the Fragrant Hills. He also inscribed a "Jianbi Xiqing" (green valley and clear creeks) plaque on the palace gate in praise of the graceful natural environment there. From the paintings, after the spring waters are directed into the wall of the palace along the creek, they join up into a lotus pond and then flow out from the other side. A few groups of small courtyards are scattered inside. Obviously, the layout of the gardens during this period was quite open and liberal.

After all, as the Fragrant Hills is quite some distance from the capital city, traveling to and fro made the emperor feel inconvenient. So he turned his eyes to Chengxin Yuan 5 km away from the Fragrant Hills and expanded the garden. In 3 years, he was already residing there every now and then. In the morning, he went to the "Front Pavilion" to meet with his ministers and discuss state affairs. In his spare time, he would go to the garden for pleasure. Why did his majesty choose to reside and work here? Perhaps because this place was far away from the bustles and hurries of the capital; perhaps the graceful environment of the springs and woods would make people smarter and wiser. In the 31st year of Kangxi's reign (1692), the garden was renamed "Jingming Yuan (Garden of Tranquility and Brightness)".

In the beginning of the Qing dynasty, the main hunting activities and some important ceremonies of the imperial family were held in the South Palace❶ outside the Forbidden City. In Kangxi period, however, the primary events outside the imperial palace compound had already been shifted to the geographically more advantaged northwestern suburbs of the capital. The Fragrant Hills and Jade Spring Hills were two of the most attractive places for these occasions.

❶ A large imperial hunting palace spanning Yuan, Ming and Qing dynasties. The ruins are primarily distributed in Daxing and Fengtai Districts of Beijing.

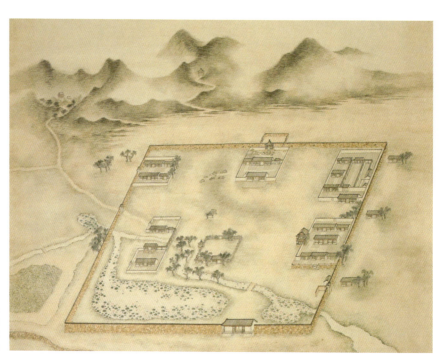

**Fig. 2.1**
Qing. Court Painter.
*A Panorama of the Fragrant Hills Palace*
(Capital Museum)

# 2.1  Jingyi Yuan

**In this section, you will get an idea about:**
1. What characterized the Inner Zone, Outer Zone, and Extra Zone scenic areas in Jingyi Yuan?
2. How did the ancient designers combine springs with garden landscaping?

One day in the 8th year of Qianlong's reign (1743), on an outing in the suburbs, Emperor Qianlong made his first visit to the Fragrant Hills Palace where Emperor Kangxi had once resided. Immediately, he fell in love with the beautiful sight there. When he saw that the plaque written by his grandfather was still hanging at the Green View Gallery, the old man's kind smile appeared before him. From then on, each time he was free in spring or autumn, he would go to visit the Fragrant Hills. He also had a grand plan in his mind—after expanding Yuanming Yuan, he would build the entire Fragrant Hills, along with the temples and palaces there, into an unrivaled upland palatial garden. This way, he would obtain the exclusive right to the thousand-year-old cultural site there.

In the first lunar month of the 11th year of Qianlong's reign (1746), the Fragrant Hills cum Jingyi Yuan (Garden of Pleasant Tranquility) was officially inaugurated. The entire area covered 1,565,000 m² of land. The name "Jingyi" is only one word's different from "Jingming", the name given by his grandfather to the Jade Spring Hills. In his *Note of Jingyi Yuan*, Emperor Qianlong described its philosophic meaning as "**Accords with the philosophic concept of Zhou Dunyi; generates some echoes to the 'origin of everything' ideology**". As to the sensitive topic of launching massive constructions, the emperor claimed that "**the garden construction was not originated by him. It was based on previous construction; the place was intended as a palace compound to show concern for the servants**". In reality, except a few large temples and sporadic imperial buildings in the garden, the majority of the scenes were designed and constructed under his auspices. (Fig.2.2-Fig.2.4)

## The formation of the "Three Zones"

In order to plan the entire Jingyi Yuan, Emperor Qianlong first enclosed the more closely distributed scenes already there into the scope of the Inner Zone with bounding

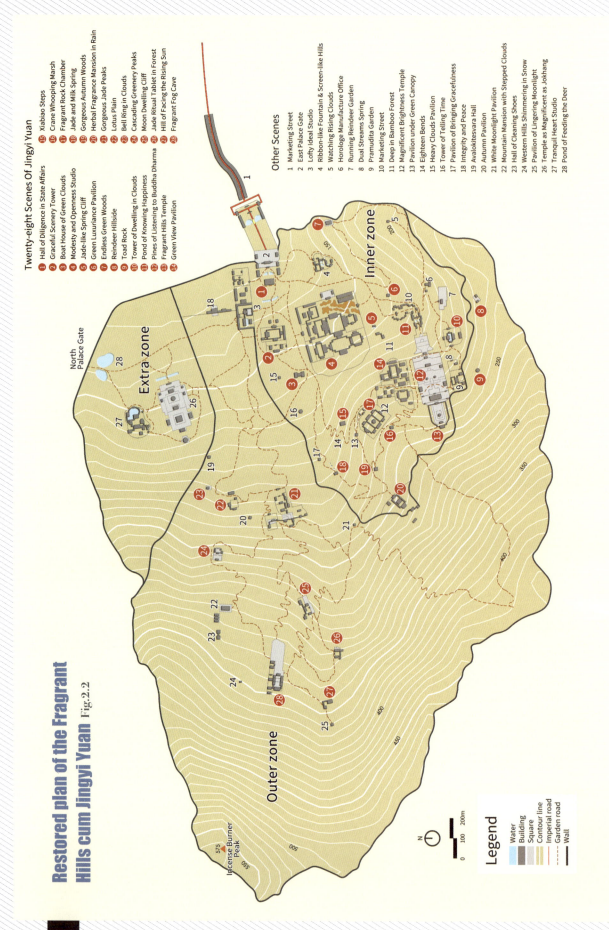

Fig. 2.2 Restored plan of the Fragrant Hills cum Jingyi Yuan

**Fig. 2.3** Qing. Court Painter. *A Panorama of Jingyi Yuan* (The First Historical Archives of China)

(Red numbers are for the Twenty-eight Scenes in Jingyi Yuan, black numbers are for the other scenes.)

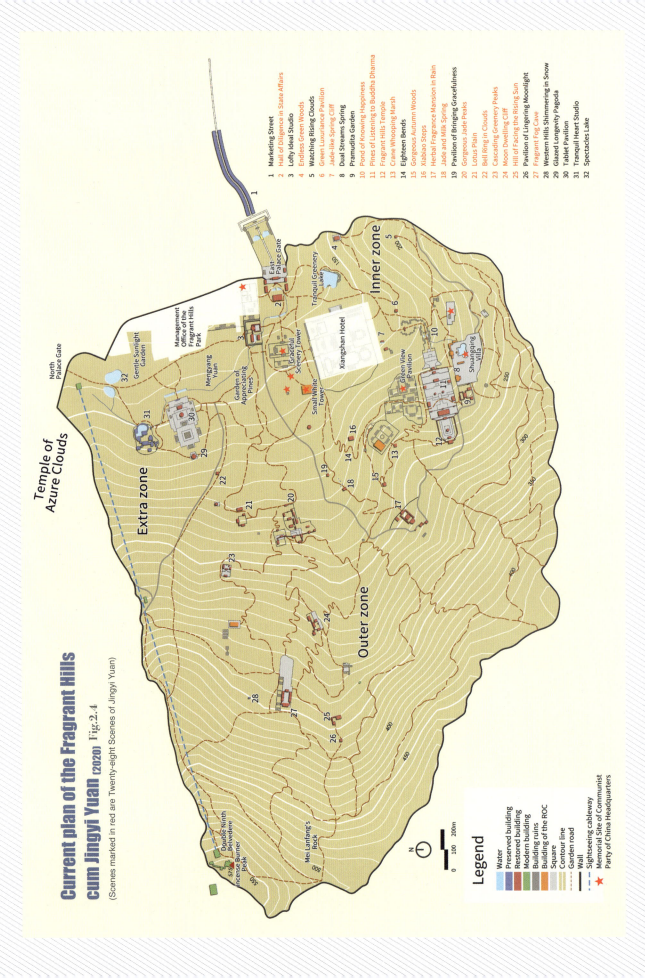

wall. The Fragrant Hills Palace, Fragrant Hills Temple, Green View Gallery, and Magnificent Brightness Temple, along with the Jade & Milk Spring and some other springs, were all included. This was the main venue for court and religious events. It covered approximately 340,000 m² of land. With woods and springs, religious buildings and living residences all in place, it was indeed a perfect site.

Next, west of the Inner Zone, in the area from the taller hill waist to the crest of the Incense-Burner Peak, he enclosed a number of locations hosting hill rocks or caves into the compass of the Outer Zone. This was the area for watching natural sceneries and taste snow water. It covered approximately 1,045,000 m² of land, which more than tripled the area of the Inner Zone. As the hills were steep and the area was vast, the buildings were not intended to be grand and tall. Instead, small courtyards were disposed in great flexibility. Winding hill paths were built to link to the relatively distant scenes. Finally, in the 34th (1769) and 45th (1780) years of Qianlong's reign, he installed an in-the-hill water yard named Moderation Hall (Tranquil Heart Studio) (Fig.2.5), and a giant buddhist temple integrating Han and Tibet characters—the Temple as Magnificent as Jokhang, at the piedmont immediately adjacent to the north of the Outer Zone. This area was called the Extra Zone. It covered approximately 180,000 m² of land. Up to then, the "Three Zones" layout in Jingyi Yuan had been finalized.

**Fig. 2.5** The water yard at the Moderation Hall

## Courts and temples in the Inner Zone scenic area

In the spring of the 11th year of Qianlong's reign (1746), the emperor hastened to the just completed Jingyi Yuan and indited poems in honor of each of the scenic spots with 28 three-word titles that he had already conceived. That marked the birth of the famous "Twenty-Eight Scenes of Jingyi Yuan". These 28 poems and the corresponding paintings have become a real record for us to interpret historical features and ideologies today.

Inside the Inner Zone scenic area, the Fragrant Hills Palace from Kangxi period was expanded into a bedchamber zone for the emperor and his concubines, the Modesty and Openness Studio. The name of this scenic spot means that only by staying open-minded can one keep an insight into everything. This is an advice to the emperor himself. The main entrance and main hall of the whole garden were no longer kept there. They had been moved northward to the regularly aligned Palace Gate of Jingyi Yuan and the Hall of Diligence in State Affairs (Fig.2.6). In explaining why he named the main hall "Diligence in State Affairs", Emperor Qianlong said that he was modeling himself on his grandfather and father, as they had set a Hall of Diligence in State Affairs on the Yingtai Islet of the West Palace and in Yuanming Yuan, respectively. So, even when he came to Jingyi Yuan for pleasure, he would not leave state affairs for delay.

The Hall of Diligence in State Affairs was shielded by a group of high-rising rockeries behind to separate the hall from other areas. Even if you walk up the winding path, you will see that the axis line of the hall is still controlling the Graceful Scenery Tower, Boat House of Green Clouds, and Tower of Telling Time on the hillside. The Boat House of Green Clouds, although named a "Fang (boat)", was not shaped like a boat, nor was it built in water like most boat houses. Instead, it was built by imitating the Cloud Sail and Moon Boat, one of the "Thirty-Six Scenes of the Imperial Mountain Summer Resort". To Emperor Qianlong, this was no different than a boathouse built in water. It would also remind him of the warning: "While water can carry a boat, it can also overturn it".

**Fig. 2.6**
Wu Xiaoping. Autumn view of the Hall of Diligence in State Affairs

**Fig. 2.7**
Backward garden of the Fragrant Hills Temple

The temples at the Inner Zone were represented by the Fragrant Hills Temple. It mirrored the artistic style of the garden temples in North China. Now the temple has been rebuilt on the 1860's ruins. It was sited in a quiet valley in the south of the Fragrant Hills. The building was seated to face the east and constructed in line with the terrain. Unlike other temples normally composed of a number of courtyards, from the bottom to the top of the hill, this temple contained a marketing street, a temple, and a garden. Emperor Qianlong expanded and enriched the scenes at the temple, enlarging its total land coverage to more than 30,000 m².

In the Qing dynasty, Marketing Streets first appeared at Changchun Yuan in Kangxi's period and at the entrance of the Sravasti City at Yuanming Yuan in Yongzheng's period. Later on, they appeared many a time in the imperial gardens like Charngchun Yuan and Qingyi Yuan. Although these special business streets contained goods-selling stores and small civilian temples, they did not serve real commercial functions. Instead, they were a scenic arrangement for immersed experience. There, imperial court people, including the emperor himself, played different roles in the marketplace to cater for the emperor's special desire to lay down his paramount class status and experience civilian life.

Behind the Marketing Street was a five-layer terrace temple and garden, with height difference exceeding 50 m.The Buddha Theophany Hall on the high terrace stands out as the tallest and grandest part of the temple. Its area is larger than 700 m². The particolored glazed tiles look extremely splendid in the sunlight, giving an illusion of a Buddhist wonderland. Yet the garden lover Emperor Qianlong did not stop there. To enhance the grandeur of the temple, turn it into a landmark of the Western Hills, and allow overlooking mountains and plains in the distance, he made a bold decision. He decided to extend the axis line of the temple further up to the top of the hill so that it would run across the gate hall—Broad Outlook Hall and the three-story Tower of Fragrant Campaka Woods❷. There, a steep climbing corridor❸ was also installed to link the tower on the summit, the Tower of Placing Relaxation on Azure Glow, to create a cliff-hanging visual shock and height-ascending experience (Fig.2.7).

❷ A flower recorded in the Buddhist scriptures. It has yellow color, strong fragrance and tall trunk.

❸ A corridor in ancient buildings constructed following the terrain.

## Natural interests and charms in the Inner Zone and Outer Zone

The Fragrant Hills Temple, as an artificial scene, represented the nature-excelling gardening and architectural arts. It showcased a distinctive political orientation or religious belief. Here in Jingyi Yuan, the most distinctive parts should be the several scenic spots at the Inner Zone and Outer Zone. They are smaller in size, but still provided for the appreciation of natural wonders.

The Gorgeous Autumn Woods was simply a self-standing pavilion. But here one could savor the full color plants like gingko and maple that were "**yellow and red in the frost of late autumn**" (the Preface to the Poem *The Gorgeous Autumn Woods*) (Fig.2.8), as well as giant rocks in the hill named "Cascading Greenery Rock" and "Rock Screen with Vine".

The Crane Whooping Marsh and Reindeer Hillside were once used to raise rare animals and birds from tributaries like "sea crane" and reindeer❹(Fig.2.9).

The Toad Rock, Jade Ritual Tablet in Forest, and Cave of Facing the Rising Sun were all grotesque peaks, weird rocks, or serene caves commonly found in the places of interest in China. The names reflect the rich imagination of the ancients.

The names "Xiabiao Steps" and "Eighteen Bends" highlighted the tortuosity and steepness of the hill paths. Only an open building was built on the tall terrace in the hill for a short break.

The Herbal Fragrance Mansion in Rain (Fig.2.10), Moon Dwelling Cliff, together with the Cave of Fragrant Fog and Study With Bamboo Tea Stove at the highest points,

❹ A rhinoplax vigil originating from East Asian tropical rain forests and a deer originating from Heilongjiang.

were themed around natural elements: rain, moon, fog and snow. They also provided viewers with the spiritual enjoyment of "**The height of man is that which he can attain; the distance that he can see is that which he can see. At that point he can be self-aware**" (the Preface to the poem *The Cave of Fragrant Fog* by Emperor Qianlong).

Besides, Jingyi Yuan is also a good example of scoping the external scenes into own field of appreciation. This is what people call "borrowing scenery from aside". For example, the Endless Green Woods, from which one could overlook the Jade Spring Hills and the capital city eastward, lay next to the bounding wall at the easternmost end of the garden. The name "Endless Green Woods" (everywhere green) comes from a famous verse of Tang poet Du Fu's *Gazing at Mount Tai*: "**What shall I say of the great peak? The ancient dukedoms are everywhere green**". It means that although here is not Mount Tai, one can still appreciate a similar artistic conception. Coincidentally, later on, Emperor Qianlong built a palace in worship of Mount Tai on the west side of the Jade Spring Hills. The Bell Ring in Clouds at the northernmost end of the Outer Zone was actually not a bell. It means that inside this small pavilion, one can listen to the bells from the temples outside the garden, such as the Temple of Azure Clouds, Temple of Universal Consciousness in Ten Directions (i.e., Temple of Reclining Buddha), and even the Temple of Awakening and Vitality (i.e., the Grand Bell Temple). The name itself is suggestive of a strong sense of zen.

**Fig. 2.8**
Qing. Court Painter. The Gorgeous Autumn Woods in the *Album of Paintings of Jingyi Yuan* (Palace Museum in Shenyang)

**Fig. 2.9**
Qing. Giuseppe Castiglione. *Scroll Painting of the Reindeer from the East Sea* (Palace Museum in Taipei)

**Fig. 2.10**
The cascading corridors at the Herbal Fragrance Mansion in Rain

## Combination of landscaping with water diversion

Nobody can live without water. Before Jingyi Yuan was built, a lot of temples had been sited there, again because of the plentiful water supply there. At the south and west side of the Inner Zone section of Jingyi Yuan, the spring holes at the Dual Streams Spring, Jade-like Spring Cliff, and Jade & Milk Spring (Fig.2.11) were the primary water suppliers for the scenic spots Pramudita Garden, Fragrant Hills Temple, Pond of Knowing Happiness, and Modesty and Openness Studio. In order to solve the water shortage in the Hall of Diligence in State Affairs section at the north side, Emperor Qianlong spared no manpower and material to install an artificial canal in the upland to divert water from the Monk's Stave Spring at the Temple of Azure Clouds more than 800 m away.

While diverting water, to make the best of spring water for landscaping, they first stored the spring water along the Temple of Azure Clouds down into the Spectacles Lake, then directed the water into the Studio of Ancient Zither Sound north of the Hall of Diligence in State Affairs, by way of the lake surface at the Tranquil Heart Studio and the Crescent Moon River at the Temple as Magnificent as Jokhang. There, the spring water produced a sound due to the falling head. This gave the building an elegant name of "Zither Sound" (Fig.2.12). Finally, the water ran past the canal and fell into the pond beside the Hall of Diligence in State Affairs in form of waterfalls. The water system flew past a number of scenic spots. Although it was essentially a water conservancy project, as an artistic treatment was used during the construction, it became an extraordinary evidence for the high garden-making expertise of the ancients. Similar approaches were also applied in the vicinity of the Fragrant Hills Temple. The Dual Streams Spring water under the Toad Rock ran past the Pramudita Garden, Tower of Dwelling in Clouds, and Pond of the Knowing Happiness. It then joined the water of the Jade-like Spring Cliff (Fig.2.13). Together, the two streams of water flew through the Ribbon-like Fountain & Screen-like Hills at the piedmont and then out of the wall of Jingyi Yuan.

**Fig. 2.11**
Qing. Court Painter. *The Jade and Milk Spring* in the *Album of Paintings of Jingyi Yuan* (Palace Museum in Shenyang)

**Fig. 2.12**
The rockery waterfall in front of the Studio of Ancient Zither Sound

**Fig. 2.13**
The rockery waterfall and Crystal Sound Pavilion at the Jade-like Spring Cliff

**Fig. 2.14**
Qing. Court Painter. Part of the *A Panorama of 27 Elders Traveling in the Fragrant Hills* (Palace Museum in Taipei)

Emperor Qianlong often came to Jingyi Yuan for short stays. According to statistics, throughout his reign, the emperor had lodged there for 72 times. His length of stay totaled around 220 days. The exceptionally advantaged terrain conditions and the beauty of woods and springs also made it the venue for two important imperial events: height ascending on the Double Ninth Festival and the "Banquets of Three Groups of Nine Elders ❺" for empress dowager's birthday celebration (Fig.2.14).

❺ A traditional event in which the emperor invited 9 civil ministers in office, 9 military ministers and 9 retired ministers to take part in the garden party.

# 2.2 Jingming Yuan

**In this section, you will get an idea about:**
1. How did the "First Spring Under Heaven" at the Jade Spring Hills take its name?
2. From which case did the design of the temples and pagodas at the Jade Spring Hills draw inspirations?

The 14th year of Qianlong's reign (1749) marked the official kickoff of an enormous water conservancy project of the capital city in the northwestern suburbs. The Jade Spring Hills cum Jingming Yuan, as one of the most important water source areas, also welcomed an opportunity for a "gorgeous upgrade". Its general designer was Emperor Qianlong, who had just completed the construction of Jingyi Yuan a few years before.

Supposedly, Jingming Yuan in Kangxi's period only included the southern waist and piedmont of the Jade Spring Hills. Its coverage was quite modest. After a close survey, Emperor Qianlong discovered that there were 6 spring holes surrounding the hill in a U-shaped array and they had gathered into several minor lake surfaces. Meanwhile, spring waters were also rushing out of the top of the hill. So, out of the desire to broaden water volume and protect the ecology, as well as his great interest in garden making and touring, the emperor decided to make a big deal about the spring holes. Besides, there were also many remains of ancient temples and caves in the hill. If massive construction of religious buildings were carried out at the same time, it would indeed be "shooting many hawks with one arrow".

In the 6th month of the 18th year of Qianlong's reign (1753), Emperor Qianlong identified the "Sixteen Scenes of Jingming Yuan" and wrote poems in praise of them (Fig.2.15). He also ordered the painters to paint a picture for each of the scenes. Although the garden was not fully completed, he had already determined the design scheme. Among the 16 scenic spots, 6 were related to spring water [6]. According to geological research, spring waters at the Jade Spring Hills are plentiful simply because here is the only outcropping spot for the underground aquifers around the Western Hills. The Jade Spring has such a large flow rate that it never dries up even in winter. It is the pivotal spring water across the THFG and even in Beijing. Inside the garden, the most representative, most water-abundant spring hole—the Jade Spring (First Spring Under Heaven)—is located at the east-facing foot in the south part of the Jade Spring Hills. It has also produced a natural lake. Continuing with the earlier layout, Emperor Qianlong used this place as the core part of Jingming Yuan. Here, he laid out 5 of the "Sixteen Scenes": the Broad Mind & Universal Justice, Lotus-Like Hill in Sunlight, Gushing Spring of Jade Spring Hills (Fig.2.16), Comprehensive Imitation of Shengyin Temple, and Fine Shade of Cloud-shaped Canopy.

[6] These included the Gushing Spring of Jade Spring Hills, Hill Chamber with Bamboo Tea Stove, Supervising Plowing by the Stream, Scenery of Tearing Silk Sound Lake, Mirror-like Water Reflecting the Sky, and Spring of Zither Sound in Canyon.

**Restored plan of the Jade Spring Hills cum Jingming Yuan** Fig.2.15

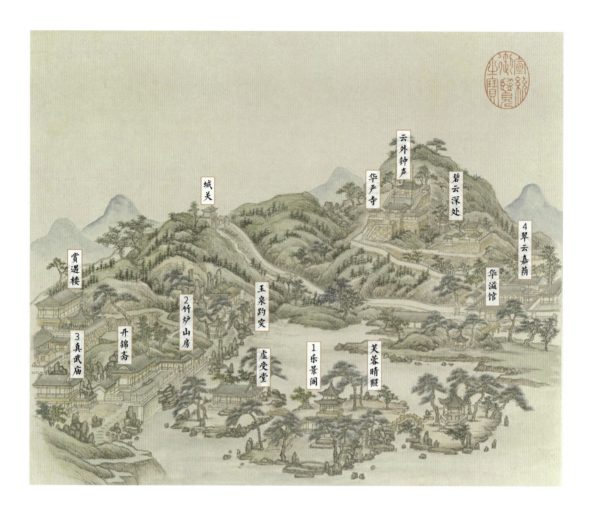

## Creating scenes around spring water

As an imperial garden, following the former practice of Changchun Yuan and Jingyi Yuan, the first things present in Jingming Yuan were a regularly-aligned palace gate and an office area. So Emperor Qianlong placed the palace entrance, interior and exterior reception houses for officials, and the main hall in the garden along an NS-trending axis line. The name of the main hall, Broad Mind & Universal Justice, comes from the famous quotation of Cheng Hao (1032—1085): "**The scholarly attainment of a man of noble character is no better than being liberal and evenhanded and complying with the trend of nature**". The main hall had the same name as one of the Forty Scenes of Yuanming Yuan. Through a length of corridor, it was combined with the rear hall "Embrace All Nature" into a quadrangle dwelling. This enabled the emperor to meet with his ministers from outside and appreciate the woods and springs inside.

In the Jade Spring Lake, there was a group of three islands named Lotus-Like Hill in Sunlight, including one big island and two smaller ones. Although named so, the place had nothing to do with lotus at all. Nor was it a continuation of the former name "Lotus Palace" built by Emperor Wanyan Jing of the Jin dynasty. It was so named simply because in the eyes of Emperor Qianlong, the Jade Spring Hills looked just like a green lotus blossom in the light of the sun. The 3 islands symbolized the 3 celestial mountains in the Chinese myths—Penglai, Fangzhang and Yingzhou. Their arrangement resembled that of the Immortal's Residence on Penglai Island at Yuanming Yuan. The larger island in the middle was the symmetrically-aligned Belvedere of Enjoying Views (1# in Fig.2.16) ; the other two by its sides were simply a pavilion and a water pavilion.

As the Jade Spring Lake was very close to the hill body and the flat space there was limited, the group of buildings, namely, the Hill Chamber with Bamboo Tea Stove (2# in Fig.2.16) imitating a teahouse at Mount Hui of Wuxi, the Comprehensive Imitation of Shengyin Temple imitating the travel palace at Shengyin Temple by the West Lake of Hangzhou, and the Zhenwu Emperor Temple (3# in Fig.2.16) built in worship of Zhenwu God, was aligned in a compact beeline along the western bank of the Jade Spring Lake (Fig.2.17).Emperor Qianlong placed his bedchamber for short breaks in a south-facing dwelling compound on the northern bank. He named the place "Fine Shade of Cloud-shaped Canopy" (4# in Fig.2.16) after two ancient cypress trees believed to have been planted in the Jin dynasty.

In order to better appreciate the other spring and lake views at the foothill, from northeast to southeast, smaller mixtures of boathouses, water pavilions, pavilions and corridors of varying shapes were also deployed next to the poetic spring holes of the Precious Pearl Spring, Gushing Jade Spring, Spring of Testing Ink and Spring of Tearing Silk Sound, as well as on the banks of a few slender lake surfaces. Like what were found in the Kangxi-named Crystal Sound Belvedere, they also had visually and auditorily suggestive names like Extensive Green Hall, Gazebo of Rinsing Faraway Green and Corridor of Zither Sound. To the west of the hill, the agriculture-oriented Emperor Qianlong opened two tracts of paddy fields covering about 8,000 m² of land. The fields were irrigated with water from the Spring of Splashing Pearls and named "Supervising Plowing by the Stream" (Fig.2.18). This way, all the springs could lend a hand in garden-landscaping before they flew into the farmlands or lakes of water storages. —This was much the same as the layout of Jingyi Yuan.

China has a long history of tea culture. The ancients believed that making tea with spring water pleases both the tongue and the mind. Hence, in order to pursue a higher realm, they often judged how good a spring water was by whether its taste was sweet, whether its mass was light, and whether it helped extend people's longevity. They also came to the conclusion that "**the lighter the mass, the sweeter the taste**".

**Fig. 2.16**
Qing. Court Painter. Gushing Spring of Jade Spring Hills in the *Album of Paintings of the Eight Views of Yenching* (Palace Museum)

**Fig. 2.17**
Old photograph of the western bank of the Jade Spring Lake

**Fig. 2.18**
Old photograph of the Supervising Plowing by the Stream and Rainbow Bridge (ROC period)

御製玉泉山天下第一泉記

玉泉山天下第一泉記

水之德在養人其味貴甘其質貴輕

然三者正相資質輕者味必甘飲之而

蠲痾益壽故辨水者恒於其質之輕

重分泉之高下焉嘗製銀斗較之京

2.19 2.20 2.21

2.22

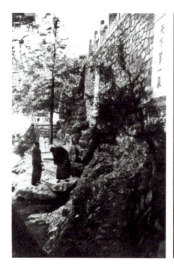

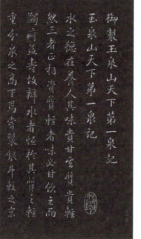

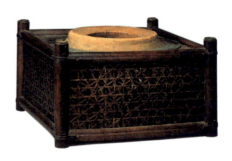

Toward such a famous spring in his own palatial garden, it was just natural that Emperor Qianlong would make a careful judgment. He sent his men to all famous springs in his empire to weigh the water with the same silver measuring bucket. Sure enough, except inferior to the snow water from Beijing's Western Hills, the spring water from the Jade Spring Hills was the lightest in the country. Then, the honorary tile the "First Spring Under Heaven" would belong to none but the Jade Spring water (Fig.2.19-Fig.2.20). So, the Hill Chamber with Bamboo Tea Stove not far away from the Jade Spring became the special venue for his tea drinking. For this reason, he also told people to make bamboo tea stoves (Fig.2.21) typical of Wuxi to experience the graceful lifestyle of the literati in regions south of the Yangtze River.

Besides enjoying tea, the emperor also did two special things. First, he discovered on this tour that, including himself, people not having truly seen the Jade Spring would praise this scene as the "Hanging Rainbow" of Jade Spring Hills, one of the "Eight Scenes of Yenching". The term Hanging Rainbow means that the spring water flows all the way down like a waterfall. But the fact is: the water gushes from underground. Hence, in order to prevent people from passing down the same wrong observation to later generations, he renamed the scene "Gushing Spring of Jade Spring Hills" and inscribed this name into a stone tablet as a declaration (although civilians were forbidden to enter here). Besides, the emperor also believed that the constant outflow of water from the Jade Spring Hills to benefit his country and people was undoubtedly a blessing by the heaven. So, to thank the gods for offering the "**water of unparalleled merits and virtue**", he built a Dragon King Temple on the high terrace next to the spring hole and frequently went there for worship.

Besides springs gushing out at the bottom of the hill, things were no different at the top. There, a courtyard specifically built for watching the hilltop springs was seated in the valley between the two peaks in the middle and south. This was the Spring of Zither Sound in Canyon (Fig.2.22). Its layout resembled the scene Lush Maple on Green Islet at the Imperial Mountain Summer Resort. The name of this scene describes both its location of a gorge and the sound of music produced by the spring water running past. In reality, this was also a venue for looking east into Qingyi Yuan and the paddy fields. The open houses

**Fig. 2.19**
Old photograph of The *First Spring Under Heaven* (ROC period)

**Fig. 2.20**
Part of the *Note of the First Spring Under Heaven* by Emperor Qianlong

**Fig. 2.21**
Bamboo Stove of the Qing dynasty (Palace Museum)

**Fig. 2.22**
Wu Xiaoping. The Spring of Zither Sound in Canyon After Snow

on the east side precisely confirmed this design intention.

Apart from the springs inside the garden, the spring waters from the Western Hills were also used in the landscaping. In the northwest part, the Ripples Studio, a garden-in-garden, was specifically designed for watching the waterfalls and lake views. This compound included scenic spots immediately adjacent to waterfalls, such as the Pavilion of Waterfall-Hanging Eave and Rapid Waterfall Belvedere, and lake-front buildings like the Natural Interest of Streams and Hills. The placement of the buildings made the scenes in the garden even more fascinating. Now, they were pleasant not only to the eyes, but also to the ears.

## Building temples in the light of local conditions

The originally existent Avatamsa Temple and Avatamsa Cave were located at the waist of the hill. The architectural form of the temple was fairly monotonous. If viewed from the eastern bank of the newly opened Kunming Lake, there was nothing special of the Jade Spring Hills except the hill body itself. On his trip in the south, Emperor Qianlong was deeply impressed by the Golden Hill Temple of Zhenjiang. That temple and the lofty Pagoda of Mercy & Longevity stood on the hillside on the southern bank of the Yangtze river. Together, they take on the artistic feature of "hills wrapped by temples", "hills heightened by pagodas". Wasn't it a good idea to borrow the architectural idea of the Golden Hill Temple into the design of the imperial temples in Beijing❼? So, in the 24th year of Qianlong's reign (1759), to the east and south of Jingming Yuan, the Lake of High-level Water, a large lake surface covering more than 500,000 m² of land, was completed at the same time as the upland Fragrant Rock Temple. To the north of the temple also stood a 7-story, octagonal pagoda. It was as tall as 50 m.

❼ the Grand View in Heaven scene at the Imperial Mountain Summer Resort also followed the same principle by Emperor Kangxi.

[8] Following this war, Burma became a vassal state to the imperial court. This submissive relationship ended in the mid 19th century with the invasion of the western powers.

On the immense lake surface rippled the inverted reflections of the Jade Spring Hills and the pagoda, producing equally intriguing effects as those in Zhenjiang. They were an excellent piece of imitative landscaping. The Tower of Reflection in Lake in the middle of the lake, as the opposite scene to the entire Jingming Yuan, was accessible by boat to get a full view of the Jade Spring Hills. Meanwhile, the Dipankara Pagoda seated close to the summit was undoubtedly a new landmark building for the entire THFG. The great sight produced by this pagoda could be viewed whether from Qingyi Yuan to its east, from Haidian, or from Jingyi Yuan to its west... In the beginning, Emperor Qianlong had thought about building it into a "**nine-story pagoda**" (The Preface to the *Shadow of Pagoda at the Jade Spring Hills*). Unfortunately, discouraged by an unexpected accident to the Longevity Pagoda at the Longevity Hill, he had to decide to cut two stories off the design (see more details in Chapter 4). Still, the pagoda was subsequently proved to be a finishing touch for its overall visual beauty and robustness. Besides, the 3 temple buildings erected on the hillside—the Deep in the Azure Clouds, Fragrant Clouds and Dharma Rain, and Distant Bell Ring in Clouds (Avatamsa Temple)—also set off the overall grandeur of the Fragrant Rock Temple (Fig.2.23).

In the 36th year of Qianlong's reign (1771), a hybrid Han-Burma-style combination of temple and pagoda appeared on the northern peak of the Jade Spring Hills—the Sumeru Temple and Sumeru Pagoda. They were memorials in honor of the expedition against the chieftain rebellion in Mubang, Burma [8]. Although the Sumeru Temple consisted of only one courtyard compound, it helped create a relatively balanced composition for the sight of the Jade Spring Hills: one pagoda on each of the northern and southern peaks. At the northern tip of the temple, there was a Pavilion of Leaning on Clouds. From there one could overlook the Tibetan-style blockhouses complex at the Small Western Hills in the distance.

Emperor Qianlong correlated the all-moistening Jade Spring water with the image-symbolizing Mount Tai. He believed that they play the same great role. But Mount Tai is after all "**hundreds of kilometers away from the capital city**". It was not so convenient even if he wanted to thank the gods there. So he erected a "**regularly shaped and magnificently structured**" temple palace on the west side of the Jade Spring Hills. That was the Benevolent Cultivation Palace (Fig.2.24). The palace covered approximately 20,000 m² of land. It was used to enshrine the God of Mount Tai and the Jade Emperor.

**Fig. 2.23**
Temples in the south of the Jade Spring Hills

**Fig. 2.24**
Sidney D. Gamble. Old photograph of the ruins of Benevolent Cultivation Palace (ROC period)

# 2.3 Eight Scenes of Yenching

Starting from Liao and Jin dynasties, the cultural deposits of the Fragrant Hills and Jade Spring Hills have become even more profound under the relentless effort of the emperors, bigwigs and large teams of literati. Earlier, the two hills were embellished with sporadic temples and travel palaces. In the heydays of the Qing dynasty, the gardens had even encompassed the entire hills and were solely intended for the imperial family. Uniquely, although they were both erected on the hillside, their scenes were far beyond the realm of natural hills and waters. They looked more like a "fully-packaged" place of interest. Scenes of all sizes, gods and Buddhas of all sects, plaques and couplets, inscribed verses and articles, furnishings and collections had all found the right place there. A good diversity of political, worship and touring events were held one after another... This represents the Chinese ancients' way of appreciating and modifying nature. It also showed how they carried the Chinese landscape culture to its extreme.

The "Eight Scenes of Yenching" originated from the poetic interest of the Jin emperor Wanyan Jing who loved travelling in the nature. After the incessant praises and chanting of the Yuan, Ming and Qing dynasty literati, including the generations of emperors, they were officially finalized in the 16th year of Qianlong's reign (1751). Emperor Qianlong even identified their exact locations and the so-called "official interpretations" with poetry inscriptions made by the emperor's order (Fig.2.25).

From the "Western Hills Covered by Snow", "Western Hills after Snow" to the "Western Hills Shimmering in Snow" (Fig.2.26), the 3 phases with only one word's difference are describing different sights of snow (some assume that they are the sights of apricot blossoms all across the hill). From the "Hanging Rainbow of Jade Spring Hills" to the "Gushing Spring of Jade Spring Hills", both phrases are depicting the morphology of spring water, yet they bring totally different impressions. In Emperor Qianlong's mind, the "Eight Scenes", having been circulated for hundreds of years, were a cultural symbol unique to Beijing. So it was highly necessary to mark them with the footprint of Qianlong era. As to the snow-covered landscape of the Western Hills, perhaps it was because the hill is the brightest when the sun shines upon its snow cover; perhaps it was because the thawed winter snow moistens everything and incubates the spring; perhaps it was because only after the snow has stopped can people fetch the snow to boil tea and enjoy the rare pleasure enjoyed by literati; perhaps it was simply because this name is more pleasant to hear... As to the Jade Spring, even if he were the only person in the world to see it, he would not allow the wrong story to be passed down anymore, although he had once made the same mistake, too; although this would leave himself being laughed at.

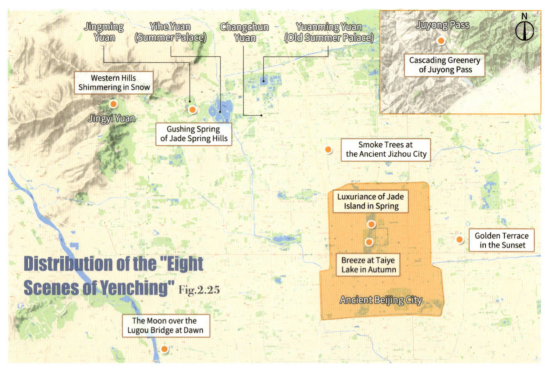

**Fig. 2.25** Distribution of the "Eight Scenes of Yenching"

**Fig. 2.26** Qing. Court Painter. The Western Hills Shimmering in Snow in the *Album of Paintings of the Eight Views of Yenching* (Palace Museum)

### Yuquanbaotu (The Gushing Spring of Jade Spring Hills)
### Aisin Guaro Hongli, 1751

All springs of the Western Hills flow deep underground. When they reach the gap in the middle of the hill, the water gushes out, rolling up and off like snow and waves. Even the Spouting Spring in Jinan is no better than it here. People who wrote poems in praise of the Eight Scenes saw it as a hanging rainbow, yet this neglects the true essence of this scene. In order to make a rectification and declare its identity as the best spring around the world, I hereby write a poem as record.

*The Jade Spring is said to drop here like a rainbow, its trueness is really dubious.*
*Gusting and rolling for thousands of years, when has it ever dropped from high above?*
*In the open pond shines the milky moon, on the inverted rock fly the pink petals.*
*Credulous I may sound, as I also followed the suit and spread the false legend.*

### Xishanqingxue (The Western Hills Shimmering in Snow)
### Aisin Guaro Hongli, 1751

The ridges and peaks of the Western Hills stand layers upon layers and it is hardly possible to name each of them. As they are in the west of the capital city, they are collectively called the Western Hills. The high terrain and low temperature is very welcoming to accumulation of snow, making the hills look like a sharpened jade. Now that a Jingyi Yuan has been built on the Fragrant Hills, so a mark is put up at the ridge as an identification.

*A place of interest was recorded long ago, where the great hills and rocks hardly suggest of Beijing.*
*A heavy snow falls just at the right time, offering a fascinating scenery at the moment it stops.*
*Around the trees arise the cooking smoke, across the yard rings the temple bell.*
*An adobe is newly built on the Fragrant Hills, just to collect and brew the Sanqing tea.*

# Touring the Three Hills and Five Gardens

Landscape, Art and Life of
Chinese Imperial Gardens

# Chapter 3
# Changchun Yuan and Yuanming Yuan
## — The Heart of the Empire amid Artificial Hills and Waters

# 3.1 Changchun Yuan

**In this section, you will get an idea about:**
1. How did Emperor Kangxi try to be thrifty when building Changchun Yuan?
2. Why was Changchun Yuan a publishing house, a museum, an experimental field, a zoo and a botanic garden?

## From a private garden to the heart of empire outside the Forbidden City

In the early years of the Qing dynasty, after Emperor Kangxi suppressed the Revolt of the Three Feudatories and recovered Taiwan, the political situation became stable (Fig.3.1). In the 1684 (23rd year of Kangxi's reign), the disease-ridden emperor took a fancy to the former site of Qinghua Yuan in the western suburbs. He hoped to relax himself in the garden. Emperor Kangxi liked this place for two reasons. First, it was well accessible, just about 7 km's walk along the Imperial Stone Road from Xizhi Gate. Second, it had some construction basis. So, after a slight shrinkage of the size of the site, he used 3 years to build the first imperial detached palace of the Qing dynasty—Changchun Yuan (Garden of Smooth Spring). This way, the national political center was formally shifted to the THFG region. From then on, the emperor was constantly lodged there, bringing major development opportunities to this area. Hence, the important historical position of Changchun Yuan was one and only among the "Five Gardens". Immediately after that, the emperor set about building the West Garden for his young princes, an accessory garden to Changchun Yuan.

When building the imperial gardens, Emperor Kangxi tried to get everything as rustic as possible. He did so out of the consideration of saving money and avoiding criticisms. The hills and waters and some of the plants in Changchun Yuan were remnants from Qinghua Yuan of Ming dynasty. As he noted in his *Note of Changchun Yuan*: "**I told the Imperial Household Department to make some rearrangements about the old garden, using the original raised portions as hills and low-lying portions as water pools**." From the result, the two gardens did have quite a lot of similarities. They both provided rich water tour experience, a large number of rockeries, and a good variety of plants. Even so, from a private garden to an imperial garden representative of the country's image, the moulding of Changchun Yuan was still impregnated with a lot of the emperor's thoughts. As he was to

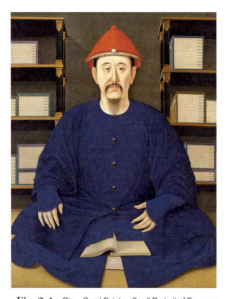

**Fig. 3.1** Qing. Court Painter. *Scroll Portrait of Emperor Kangxi Reading Books* (Palace Museum)

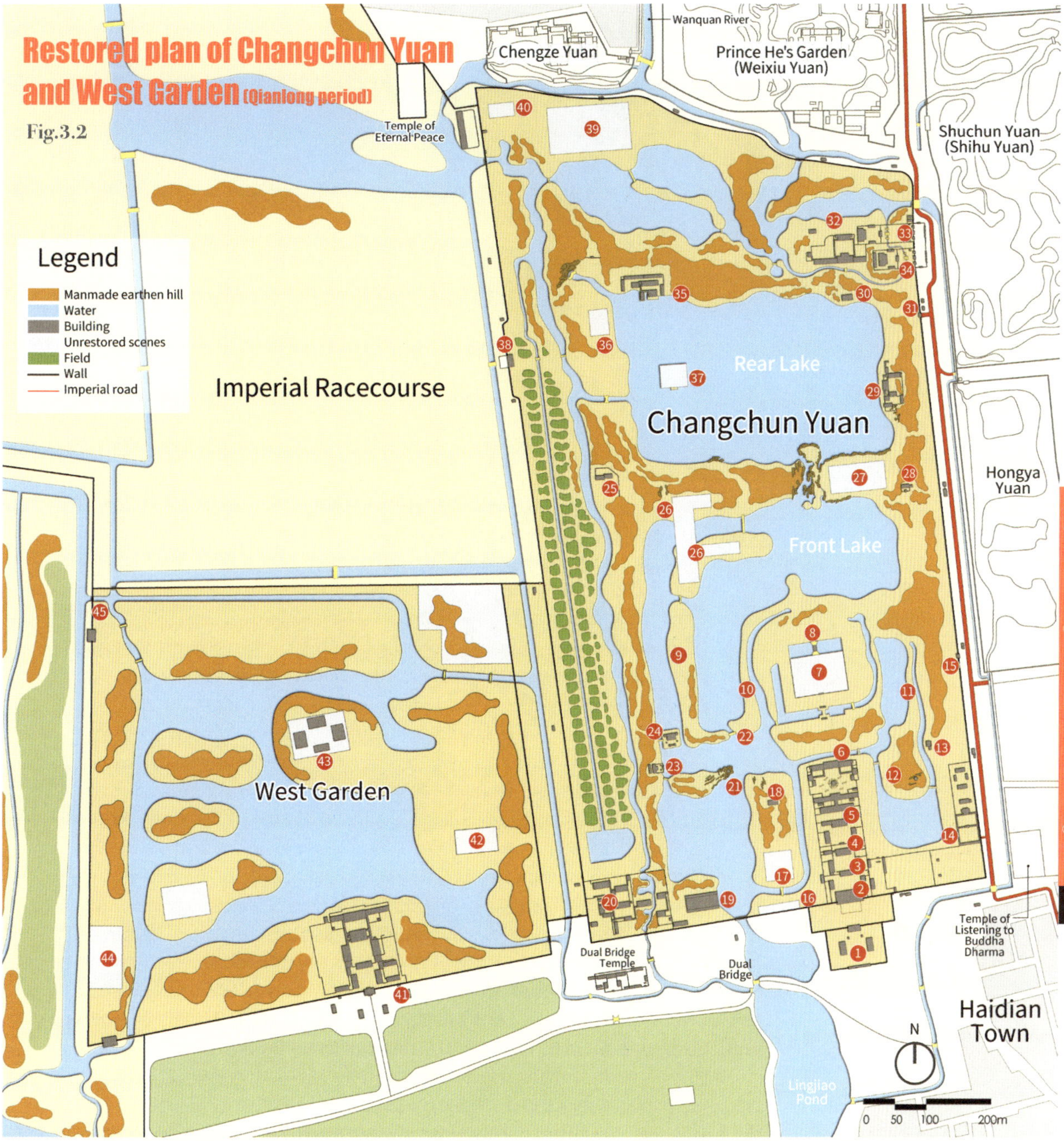

**Restored plan of Changchun Yuan and West Garden (Qianlong period)**

Fig.3.2

Legend
- Manmade earthen hill
- Water
- Building
- Unrestored scenes
- Field
- Wall
- Imperial road

Chengze Yuan
Wanquan River
Prince He's Garden (Weixiu Yuan)
Temple of Eternal Peace
Shuchun Yuan (Shihu Yuan)
Imperial Racecourse
Changchun Yuan
Rear Lake
Front Lake
Hongya Yuan
West Garden
Dual Bridge Temple
Dual Bridge
Temple of Listening to Buddha Dharma
Haidian Town
Lingjiao Pond

0 50 100 200m

1. Main Palace Gate
2. Hall of Nine Classics & Three Events
3. Secondary Palace Gate
4. Hall of Kindness as Spring Sun
5. Hall of Day Lily with Longevity & Eternal Youth
6. Mansion at the Edge of Clouds
7. Auspicious View Pavilion
8. Pavilion of Birds Flying & Fish Jumping
9. Peach Blossom Causeway
10. Fairy-herb & Orchid Causeway
11. Lilac Causeway
12. Vast and Hazy Pavilion
13. Dragon King Temple
14. Simplicity & Tranquility Hall
15. Main East Gate
16. Marketing Street
17. Studio of Appreciating Blossoms
18. Charming Pines Pavilion
19. Dock
20. Studio of Working without Ease
21. Lotus Rock
22. Pine and Cypress Sluice
23. Master Guanyu Temple
24. Bixiayuanjun Goddess Temple
25. Fragrant Blossoms Villa
26. Hall of Retaining Spring
27. Profound Literature Studio
28. Fujun Temple
29. Scattered Peaks
30. Trueness & Innocence
31. Secondary East Gate
32. Book House Surrounded by Clear Stream
33. Temple of Gratitude & Blessing
34. Temple of Gratitude & Yearning
35. Gazebo of Observing Billows
36. Pavilion of Gathering Phoenix
37. Ruizhu Yard
38. Main West Gate
39. Long Tower
40. East Book House
41. Palace Gate
42. Four Dwelling Palaces for Princes
43. Book House of Tracing the Origin
44. Pavilion of Collecting Dew
45. Main North Gate

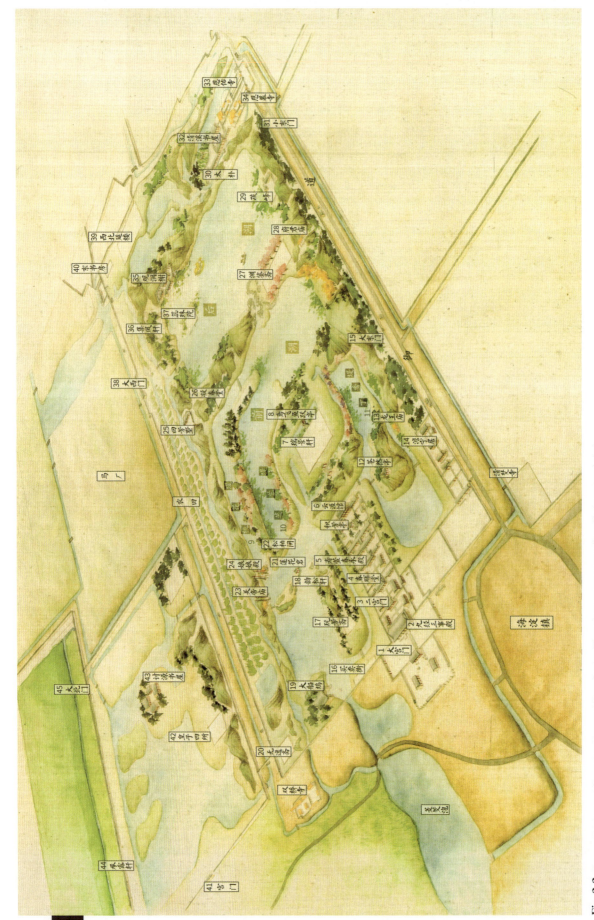

**Fig. 3.3** Restored birdview of Changchun Yuan and West Garden (The nubers are corresponding with the restored plan)

reside there for a long time, Emperor Kangxi tried to implant into Changchun Yuan all possible activities he might incur and those that would not be able to take place back in the imperial palace. These activities included:

—Dealing with state affairs, compiling books, bringing up his sons, providing for his mother and performing religious worships,

—Relaxing himself, watching crops and inspecting archery contests.

Hence, this 880,000 m² garden, with water covering more than a half of the area, became the new imperial court of the Qing Empire. Compared with the neatly aligned, magnificently-poised Forbidden City, the new garden was of a totally different style. But one thing was obvious: The owner of the garden was keener on the idyllic life inside the garden. The ancients tended to divide a garden horizontally along the eastern, central, and western axis lines. Yet the layout of Changchun Yuan was oriented longitudinally: its court, front lake, rear lake and north lake sections were disposed from south to north. The farming section stripped on the west side. The West Garden adjacent to Changchun Yuan was intended for the princes to live and study. On the northwest side was the spacious Imperial Racecourse for inspecting the troops' horse-riding and archery (Fig.3.2-Fig3.3).

## What does "Changchun (Smooth Spring)" mean?

Ancient Chinese literati were very particular about naming gardens and buildings. This was even more true for the imperial family. Then, why was this newly built garden named Changchun Yuan? Emperor Kangxi gave his answer in an article. He observed that garden scenes do not necessarily mean to be particularly suitable for viewing in spring. According to the "cycle of three-rulings" proposition of Western Han thinker Dong Zhongshu (179—104 B.C.), there are 3 springs in a year: the 11th month of the year, which is the spring under Heaven's reign; the 12th month of the year, which is the spring under Earth's rule; and the first month of the year, which is the spring under Human's rule. Also, according to the *Book of Changes*, the Qi of Origin and Nature (Qian Yuan) runs through the process of Heaven, then the four virtues of Origin (Yuan, stand for spring), Communication (Heng, stand for summer), Benefit (Li, stand for autumn), and Justness (Zhen, stand for winter) will all be fulfilled. This way, all four seasons of the year will be "spring".

From this, Emperor Kangxi was hoping that at a time of peace and order, people would each be properly provided for, enjoy their lives and remain happy; that all directions of wind and all climate changes would come and go smoothly. Obviously, Emperor Kangxi, equipped with profound deposits of the Han culture, conveyed his wishes for a beautiful world into the name of this garden. This has far greater cultural connotation than the Qinghua Yuan of the Ming dynasty.

The imperial court section was located at the southernmost end and close to Haidian Town. It mainly stretched along a 500 m long central axis. Outside, a narrow rivulet separated Haidian Town from the forbidden imperial space. Starting from the outermost Main Palace Gate and Big Screen Wall, there spread more than 10 palatial buildings of varying shapes in 8 courtyards❶. These included the Hall of Nine Classics & Three Events (the main hall), Hall of Kindness as Spring Sun, Hall of Day Lily with Longevity

❶ A normal quadrangle dwelling for ordinary residents in Beijing consists of only one or two courtyards, contrasting sharply to the imperial palace.

❷ Both the Hall of Kindness as Spring Sun and Hall of Day Lily with Longevity & Eternal Youth were names in Qianlong period. Their earlier names are unknown yet.

❸ A species of garden rockery that looks like the officials appearing before the emperor holding a jade tablet in their hands. Similar rocks are also found behind the Hall of Justness and Honesty at Yuanming Yuan.

& Eternal Youth, Mansion at the Edge of Clouds, Auspicious View Pavilion, and Tower of Lingering Freshness (possibly a remnant of Qinghua Yuan). The buildings were intended for state ceremonies and the daily life of the empress dowager❷. But the emperor did not normally work there. His office was a smaller courtyard compound located east of the central axis and close to the bounding wall. The name of the place was very "low-profile", too. It was named the "Simplicity and Tranquility Hall" after a famous quotation of the Eastern Han politician Zhuge Liang (181–234): "**One can't show high ideals without simple living; one can't have lofty aspirations without a peaceful state of mind**". The emperor chose this place as his office site possibly out of two reasons. First, the empress dowager possibly once resided in the building complex along the central axis. Second, this would make it easier for him to meet with his ministers who came in from outside the east gate. Although the location was out of the way, the environment was quite special. Its west side was close to the lake surface. On the hillside to its north were the Swordlike Rocks❸ and 2 pavilions.

Most satisfactory to Emperor Kangxi were probably the 3 causeways he placed in the Front Lake. From west to east, they were the Peach Blossom Causeway, Fairy-herb & Orchid Causeway, and Lilac Causeway. They were all named after plants. Peach trees are often planted on water banks together with willow trees to create a strong contrast of color. Lilac gives off a strong sweet smell in spring and looks purple or white in color. Fairy-herbs embody a long life or noble character. They probably referred to aquatic plants like calamus on the banks of water. Obviously, from vision and smell to cultural implications, the three earth causeways were good evidence of the emperor's artistic taste. Later on, the emperor named the long causeway at the Imperial Mountain Summer Resort "Fairy-herb Path and Cloud Causeway". The plane shape of this causeway bore some resemblance to fairy herbs.

In the Front Lake section, there was an important cultural center, the Profound Literature Studio (*Yuanjian Zhai*). This place was located between the Front Lake and Rear Lake. There, Emperor Kangxi once organized his civilian officials to compile many important books. He himself also prefaced the books. Examples included the *Album of Paintings of Plowing and Weaving* (*Peiwen Version)* in the 35th year (1696), *Collection of Ancient Painting and Calligraphy* (*Peiwen Version*), and *Comprehensive Flora* (*Peiwen Version*) in the 47th year (1708); the *Ancient Literature Collection* (*Yuanjian Version*) *and Literature Dictionary* (*Yuanjian Version*) (Fig.3.4) in the 49th year (1710); and the *Rhyme Dictionary* (*Peiwen Version*) in the 50th year (1711) of Kangxi's reign... These precious classic books bearing wordings like "Yuanjian" or "Peiwen" are important evidence for the civil administration of Emperor Kangxi. They also carry the footprints of Changchun Yuan.

The Rear Lake section of Changchun Yuan was composed of a 300 m long, 200 m wide lake surface as the main scene. Around the lake were various scenic spots placed against earth mounds, making the scenes quite private. Here was where the emperor resided and toured for pleasure. Hence, "front" and "rear" here do not only mean an orientation difference, but also the "external" and "internal" usages. Along the western, northern, and eastern banks of the lake stood a few minor buildings: the Pavilion of Gathering Phoenix, Gazebo of Observing the Billows, Trueness and Innocence, and Scattered Peaks. On the small island in the middle of the lake rose a tall tower, Ruizhu Yard. This tower constituted the visual focus of the entire area. It was a symbol for an immortal island in water, as well as the commanding height for overlooking the scenery at the Western Hills. The Gazebo of Observing the Billow was a courtyard compound on the northern bank of the lake. Its name means watching the ripples on

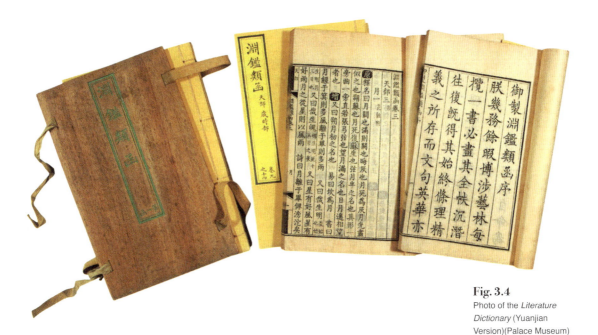

Fig. 3.4
Photo of the *Literature Dictionary* (Yuanjian Version)(Palace Museum)

the surface of the lake. The Scattered Peaks was located on the eastern bank of the lake. Here, one could look west into the undulating ridges and peaks in the distance.

The most important imperial building in this area was the bedchamber of Emperor Kangxi. Its name was "Book House Surrounded by Clear Stream". Seated in the northeast of the Rear Lake section, this large quadrangle compound was encircled by earth mounds, with the clear stream at the front and the North Lake behind. The place was definitely a low-profile yet extravagant dwelling. Yet it has been visited by some special persons around the emperor too: foreign missionaries, envoys from the Roman Vatican, and trusted ministers. According to the personal diary of Qing dynasty official Gao Shiqi (1645—1704), the emperor once picked some cherries planted there and gave them to him for a taste. This shows that the emperor and his ministers could be quite close outside the audience chamber.

The farming section was quite a special area in Changchun Yuan. In his travel notes, early Qing dynasty writer Tan Qian (1593—1658) recorded that in Qinghua Yuan, there were many plants growing in water; paddy rice was also grown there. But here in Changchun Yuan, the imperial family definitely needed no field in the garden to supply food for them. Yet the emperor still decided to retain a sizable area of fields and open a separate section for farming. That was attributable to the agriculture-oriented ruling ideology of the Qing monarchs. This area was covered by 102 experimental fields for growing vegetables, paddy rice and wheat. A 600 m long irrigation canal served to irrigate the fields. Here, Emperor Kangxi once looked over the paddy rice grown in the fields. He also recorded in his poem how delightful he felt when looking at the imperial rice in the midsummer of the 39th year of his reign (1700). The emperor also opened up a Garden of Abundant Rainfall at the West Palace next to the Forbidden City for experiencing farm work.

For the West Garden, as it was a new garden, Emperor Kangxi did not expend much energy creating its hills and waters. Hence its layout was quite simple and there were not many manual cuts and polishes inside. The garden was centered around a large expanse of lake surface. All sides of the bounding wall were encircled by earth mounds. In the south of the garden, courtyards and halls were aligned along the central axis. A peninsula stuck out of the eastern bank. On the west side of the lake rose 3 islets. Buildings were placed either on the water banks or the peninsulas or islands, rendering a very graceful environment. According to historical records, the princes were assigned to reside in 4 dwellings in the garden: the southern, eastern, central, and western dwellings. The crown prince Yinreng (1674—1725) resided in the Book House of Tracing the Origin (its location is indeterminated yet). "Tracing the Origin" means constantly pursuing the mental rule for governance and administration of state affairs. Here was a place where Emperor Kangxi's grandson, Emperor Qianlong, had often visited and contemplated. To Emperor Qianlong, when faced with problems in handling state affairs, while finding comfort from natural landscapes was helpful, it couldn't be a fundamental cure. Rather, searching for previous experiences in this quiet, secluded Book House of Tracing the Origin would probably be a wise solution. No sooner had the West Garden been completed than Emperor Kangxi began writing poems in praise of the prosperity and fragrance of the spring flowers there. He also associated the flowers with how short the human life is and told his sons to cherish time.

On the 13th day of the 11th month of the 61st year of Kangxi's reign (Dec. 20, 1722), at 19:00–21:00, Emperor Kangxi, the ruler who had initiated the last heydays of feudal age and created the first imperial detached palace in the THFG, left the world in Changchun Yuan.

## A change of fate for Changchun Yuan

In Kangxi period, Changchun Yuan was undisputedly the center of the imperial life. Later on, with the emergence of Yuanming Yuan in Yongzheng period, the position and functions of Changchun Yuan were greatly weakened. In 1843, the garden was eventually discarded. Its final boom fell in the years when it served as the abode of the empress dowager in Qianlong period. While away from the Forbidden City, the empress dowager spent most of her time here. Occasionally she would stay in Hermit Eternal Spring's Fairyland at Yuanming Yuan for short stays, too. Emperor Qianlong was a filial son. While residing in Yuanming Yuan, he would often come to visit his mother early in the morning. Sometimes he would also check how well the paddy rice was growing or watch the equestrian performance with his mother in the garden. After this, he would stay at the Pavilion of Working without Ease or the West Garden for a short time. There he would eat a meal, take a break or handle some work. In his poem, Emperor Qianlong recorded that the empress dowager enjoyed residing in the garden very much; she would prefer to stay here even when the emperor went back to the imperial palace in winter. However, in no case would she stay there until after Winter Solstice. Even though they were quite some distance apart, the son would come to visit his mother in the garden very often.

This close attachment between mother and son continued all the years until the mother died at the age of 85 (Fig.3.5). In order to pray for his mother, Emperor Qianlong, like what his grandfather had done for his great grandmother Xiaozhuang at the South Palace, built a similarly-arranged Temple of Gratitude & Yearning south of the Temple of Gratitude & Blessing (which was built by Emperor Yongzheng). Unfortunately, after the death of Empress Dowager Chongqing in the 42nd year of Qianlong's reign (1777), until the 2nd year following Emperor Daoguang's enthronement (1822), the garden had been left unused and carelessly attended for 45 years. Buildings broke down. Grass sprang up. Lakes grew silted. Although the new emperor had planned to use Changchun Yuan as his mother's abode again, he was frustrated to discover that the garden was already too dilapidated to be renovated with a small budget in a short time. So he made a hard decision to give up Changchun Yuan and use Qichun Yuan next to Yuanming Yuan as the abode of his mother. Meanwhile, in order to renovate the other gardens, the tight-budgeted Emperor Daoguang had many a time ordered to demolish all the buildings in Changchun Yuan except a few minor temples. In Xianfeng period, the two eminent ancestral temples, Temple of Gratitude & Blessing and Temple of Gratitude & Yearning, were ruined in warfare. Only their gates, the only two surface structures remained of Changchun Yuan, were left behind, recounting the few historical memories of this famous garden (Fig.3.6).

Fig. 3.5

Qing. Court Painter. Part of the *Portrait of Empress Dowager Chongqing in 80th Birthday Celebration* (Palace Museum)

Fig. 3.6

Temple of Gratitude and Blessing (right) and Temple of Gratitude and Yearning (left)

# 3.2 Clustering of gardens in Haidian

In the 46th year of Kangxi's reign (1707), all the princes were grown up. As a rule, they each should possess a separate abode and no longer reside in the same West Garden. So immediately after the Lantern Festival (the 15th of the first month), they wrote a joint letter to their father: "**The 3rd and 4th princes Yinzhi and Yinzhen and other princes are asking for your permission to build a house nearby Changchun Yuan.**" This request was approved by the emperor: "**I hereby designate the vacant lot east of the 'new garden' located north of Changchun Yuan for building your own houses.**" In the 3rd month, Yinzhi (Fig.3.7) wrote another letter to the emperor. It said that if all 7 princes, namely the 3rd, 4th, 5th, 7th, 8th, 9th, and 10th sons of the emperor, built their houses on the designated lot, the space would be far too limited. Hence as the oldest among his brothers, he would like to build his house in a farther place together with his 5th and 7th brothers.

Then the 3rd prince wrote another memorial to the emperor: "**I have bought a vacant lot southeast of Shuimo Village, adjacent to the home of Mingzhu's son Kuifang to build my house (the Xichun Yuan)**". However, we are not certain yet as to what place it was meant by "new garden". Probably it was Haoshan Yuan of the 1st prince Yinzhi. So, 7 minor gardens for the princes were added around Changchun Yuan and the West Garden. These gardens, together with Ehui Yuan (Garden of Splendorous Calyces) previously granted by the emperor to his brother Fuquan (1653—1703), Ziyi Yuan (Self-content Garden) granted to his important official Nalan Mingzhu (1635—1708), Tong Family's Garden to his important official and relative Tong Guowei (1643—1719), and Songgotu's Garden to his important official Songgotu (1636—1703), totaled 11 gardens near Changchun Yuan (Fig.3.8-Fig3.9). Garden residing had already become a common practice up to that period. Among the 11 gardens, only 6 can be located and traced for their evolution so far. They are: Ehui Yuan (not existing, part of the subsequent Qichun

Fig. 3.7
Qing. Court Painter.
*Scroll Portrait of 3rd Prince Yinzhi* (Palace Museum)

Fig.3.8 **Inferred distribution of gardens in Haidian (Kangxi period)**

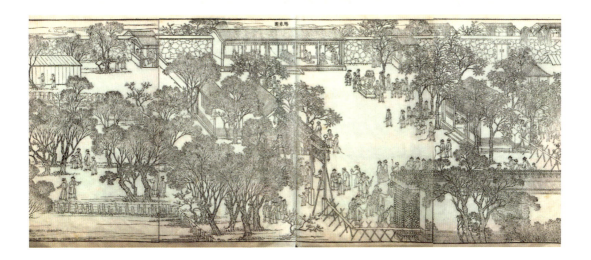

Yuan, the present-day No.101 Middle School), Haoshan Yuan (not existing, the present-day Kunming Lake), Songgotu's Garden (not existing, now the Cemetery of Northeastern-China Army), 3rd prince's Xichun Yuan (ruins, inside the Tsinghua University), 4th prince's Yuanming Yuan (ruins), and 9th prince's Caixia Yuan (ruins, the Weixiu Yuan). All the other gardens have probably vanished in the long historical evolution.

Today, people often use "Garden of Gardens" to describe the unparalleled beauty of

**Fig. 3.9**

Qing. Court Painter. Main Palace Gate area of Changchun Yuan during Emperor Kangxi's 60th birthday (Palace Museum)

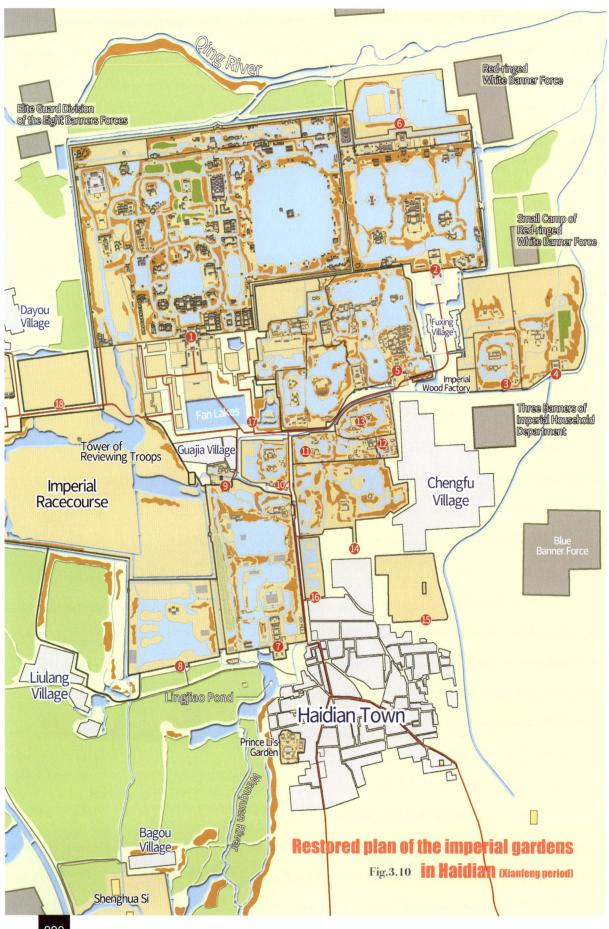

Fig.3.10 Restored plan of the imperial gardens in Haidian (Xianfeng period)

Yuanming Yuan. Nevertheless, the expression itself means that Yuanming Yuan was actually a composition of a number of gardens. If we look from regional perspective, Yuanming Yuan lived up to the name "Garden of Gardens" only when it was combined with the dozen gardens closely adjacent to it (Fig. 3.10). In Qianlong and Jiaqing periods, when the THFG was at its prime, the area in and around Yuanming Yuan was inhabited by a large number of imperial nobles and officials. They also had their mansions inside the capital. As the emperors were constantly lodged there, for the sake of convenience, they also chose to settle there.

The area enclosed by these 18 gardens covered 7,270,000 m² of land, which is about 10 folds the size of the Forbidden City. The vacant land between individual gardens also included government offices like the Imperial Opera Department, Archives Room of Yuanming Yuan and Imperial Wood Factory, and small hamlets like Guajia Village and Fuxing Village.

Besides, in Haidian Town and the surrounding hamlets, there were also many private gardens. As most of them no longer exist, the exact number is hardly countable. Such intimate distribution of bulk-sized gardens was rarely found in ancient China. Quite a large part of them are still existing today. Due to feudal hierarchy, their sizes varied vastly from 16,000 m² (Jingchun Yuan) up to 2,070,000 m² (Yuanming Yuan, the main garden).

### An inventory on the 18 imperial gardens in Haidian (Fig.3.10)

❶Yuanming Yuan, subordinated by ❷ Charngchun Yuan, ❸ Jinchun Yuan, ❹ Tsinghua Yuan, ❺Qichun Yuan, and ❻Chunxi Yuan

❼Changchun Yuan, subordinated by ❽West Garden;

These 8 gardens were encircled or connected by layers upon layers of bounding walls or bridged by overpasses. They were densely guarded by the Eight Banners troops, emphasizing their paramount authority.

Next to Yuanming Yuan and Changchun Yuan, there were a number of minor gardens with artificial hills and waters. Although their names and ownerships had undergone very complicated changes, the arrangements of the following few were quite clear:

Immediately north of Changchun Yuan were ❾ Chengze Yuan and ❿ Weixiu Yuan;

East of Changchun Yuan imperial road spread were ⓫ Minghe Yuan, ⓬ Jingchun Yuan, ⓭ Langrun Yuan, ⓮ Shuchun Yuan (Shihu Yuan), ⓯ Prince Zhi's Garden, and ⓰ Hongya Yuan (or Jixian Yuan) which served as the office venue for ministers;

Near the Grand Palace Gate of Yuanming Yuan lay ⓱ Garden of Imperial Academicians (Chenghuai Yuan) ⓲ Imperial Horse Stable (Zide Yuan).

# 3.3 Yuanming Yuan

**In this section, you will get an idea about:**
1. From which aspects can we see that Yuanming Yuan was peerless in the world?
2. How did Yuanming Yuan scale down the Chinese topography with hills and waters?
3. What typical arrangements of Changchun Yuan did Yuanming Yuan inherit?

This core of core in the garden area was Yuanming Yuan (The Perfect and Wise Man's Garden, or the Old Summer Palace), the place constantly resided by the emperor, empress and imperial concubines.

Yuanming Yuan was a large, multifunctional imperial palatial complex integrating work, residing, worship, diplomacy, as well as recreation, pleasure, military parade, and agriculture. Compared with Changchun Yuan in Kangxi period, its overall layout was completely unconstrained by the original space and financial affordability. Hence it could be created according to Emperor Yongzheng's ideal pattern. The garden was larger with more functions and better scenes. In all respects it was a "deluxe edition".

## Historical position of Yuanming Yuan

Ever since the ancient times, imperial gardens have been exceptionally advantaged in affordability and size. These are far from attainable for private gardens in the civilian world. Even the gardens of the princes in Kangxi period were no exception. However, when history selected Yinzhen to succeed to the throne, it selected Yuanming Yuan as an imperial garden. Today, we are still uncertain about the true reason why Emperor Yongzheng would not reside or work in Changchun Yuan. Whatever the reason, the newly built Yuanming Yuan did create a good opportunity for the final development of ancient Chinese gardens (Fig.3.11-Fig3.12). On the basis of the construction experience of Changchun Yuan, the new palatial garden further consolidated the garden-making style with Qing characteristics. It became an unprecedented, epic-like monumental gardening piece in history.

At its height, Yuanming Yuan, together with the accessory Charngchun Yuan and Qichun Yuan, boasted 121 thematic scenic areas (see the restored plans of the individual gardens).

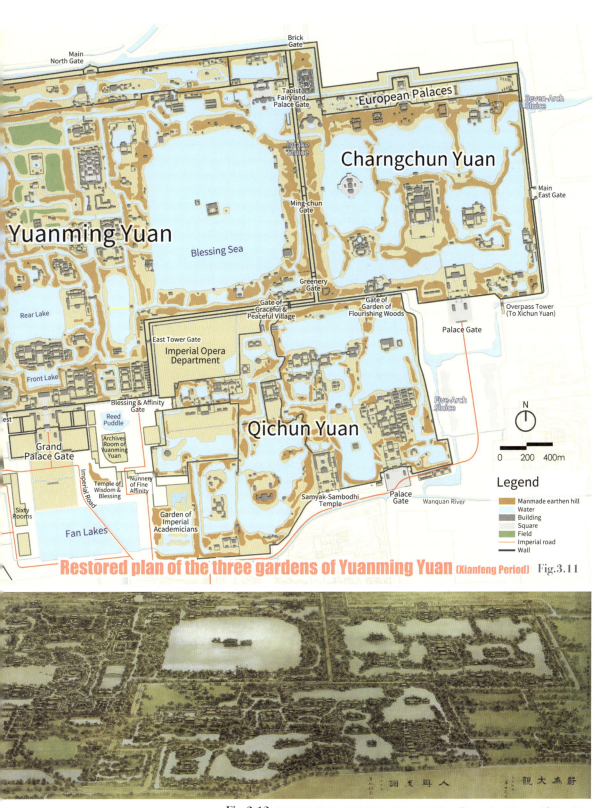

Fig. 3.11 Restored plan of the three gardens of Yuanming Yuan (Xianfeng Period)

Fig. 3.12 Bai Rixin. *Restored Birdview of Yuanming Yuan, Charngchun Yuan and Qichun Yuan*

—Yuanming Yuan was comprised of the "Forty Scenes" named by Emperor Qianlong and other 10 scenes. Its total area was 2,070,000 m² (Fig.3.13-Fig3.14).

—Charngchun Yuan was comprised of 18 Chinese-style garden scenes and one European Palaces scenic area (containing 15 scenic spots) planned under the direction of Emperor Qianlong. Its total area was 760,000 m².

—Qichun Yuan was comprised of the "Thirty Scenes" named by Emperor Jiaqing and other 22 scenes. Its total area was 700,000 m².

These scenes were themed around governing ideology, religious worship, classic literature, farmland cultivation, myths and legends, imitative construction of buildings typical of the regions south of the Yangtze River, and exotic styles. They exploited the ancients' imagination to the fullest potential. They all consisted of artificial hills and waters, palaces and halls, and animals and plants. But through organic combination, none of them were identical. This offers inexhaustible inspirations for later artists, which is perhaps the greatest artistic value in this miraculous artifact.

On a microscopic dimension, the diversely-shaped, varying-sized buildings were fitted with countless furnishings made by top craftsmen for use by the imperial family, such as wooden furniture, gold and silver ware, jade ware, porcelain ware, lacquer ware, as well as imperial collections such as centuries-old calligraphy and paintings and bronze ware. It is fair to say that these achievements were peerless before the Qing dynasty and even among palatial gardens throughout the world. Yuanming Yuan well deserves the paramount reputation. In 1792, secretary to the British mission John Barrow admitted that the sight of Yuanming Yuan was very intriguing. It was more like a perfect natural scenery rather than a mix of buildings. He believed that the scenes inside the garden were carefully designed but appeared as if they were naturally made. In 1861, in his *Letter to Captain Butler on Anglo-French Troops Expedition to China*, the great French writer Victor Hugo spoke highly of Yuanming Yuan: "**It was a kind of tremendous unknown masterpiece, glimpsed from the distance in a kind of twilight, like a silhouette of the civilization of Asia on the horizon of the civilization of Europe.**"

In early 1707, the emperor's 7 princes each built a garden for themselves. Yuanming Yuan was among these gardens. According to historical texts, in the 11th month of that year, Yinzhen invited his father to the just completed garden for a feast. That was when the name "Perfect and Wise" came into being. Later on, this plaque inscribed with the emperor-given name was hung on the Hall of Yuanming Yuan (official archives say that Yuanming Yuan was first built in 1709).

### The meaning of "Yuanming (Perfect and Wise)"

As explained by Emperor Yongzheng (i.e.,Yinzhen) in his *Note of Yuanming Yuan*, "Perfect means full moral accomplishment and appropriate behaviors; "Wise" means the ability to tell right from wrong and the wisdom of reasoning (Fig.3.15). This meaningful name conveys a father's expectations to his son. It also carried a strong political complexion, sounding quite special among the many other granted gardens.

How was Yuanming Yuan different in the early years than in its heyday? Yinzhen once named 12 of its minor scenic spots, including Peony Terrace, Goldfish Pond, Bamboo Yard, Peach Blossom Cove, Book House in Deep Willows, Vegetable Garden, and Grapes Yard. He

# Restored Plan of Yuanming Yuan (Xianfeng period) Fig.3.13

**Forty Scenes in Yuanming Yuan**

1. Justness and Honesty
2. Diligence in Government, Affection to the Virtuous
3. Peace over the Nine Prefectures
4. Carving the Moon, Tailoring the Cloud
5. Natural and Picturesque Scenery
6. Academy of Phoenix Trees
7. Mercy Cloud of Universal Blessing
8. Merged Colour of Waters and Skies
9. Winehouse of Apricot Blossoms in Spring
10. Magnanimousness of Mind
11. Inclusiveness of Ancient and Modern
12. Hermit Eternal Spring's Fairyland
13. Universal Peace and Harmony
14. Spring Peach Blossoms of Wuling
15. High Mountains and Long Rivers
16. Water-Moon Bodhimanda in Clouds
17. Great Mercy and Eternal Blessing
18. Academy of Gathering Talents
19. Vairocana's Magnificent Residence
20. Simplicity and Tranquility in Mind
21. Paddy Fragrance over the Water
22. Clear Water and Flourishing Woods
23. Mr. Lianxi's Land of Happiness
24. Spacious Paddies as Clouds
25. Fish Jumping and Birds Flying
26. North Distant Mountain Village
27. Graceful Scenery of the Western Peaks
28. Book House Appropriate for Four Seasons
29. Taoist Wonderland on Fanghu Island
30. Cleanse both Physically and Morally
31. Autumn Moon over the Calm Lake
32. Immortal's Residence on Penglai Island
33. Hill House with Graceful Scenery
34. Concealed Beauty in the Fairy Caves
35. Zither-like Water Sound over Two Lakes
36. Mirror-like Lake Reflecting the Sky
37. Broad Mind and Universal Justice
38. Sitting on Rocks by Stream
39. Distillery and Wind-blown Lotus
40. The Depths of Fairy Caves

**Other Scenes**

41. Hill House of the Purple Jade Palace
42. Conform to the Nature of Plants
43. Imitated Sails Belvedere
44. Master Guanyu Temple
45. The Vast and Transparent Sky
46. Algae Garden
47. General Liumeng Temple
48. Temple of Gathered All Spring
49. Pavilion of Literary Fountainhead
50. Sravasti City
51. Marketing Street
52. Garden of Shared Happiness

**Legend**
- Manmade earthen hill
- Water
- Building
- Field
- Contour line
- Wall

Fig. 3.14  Bai Rixin. Yuanming Yuan in the *Restored Birdview of Yuanming Yuan, Charngchun Yuan, and Qichun Yuan* (The numbers are corresponding with the restored plan)

also indited 12 poems in praise of them (Fig.3.16). We can imagine how special and different the garden was from the traditional stereotypes of gardens. Vegetables and grapes were also grown like those in Changchun Yuan. The garden was not in the least the sumptuous appearance depicted by the *Album of Paintings and Poems of Forty-Scenes of Yuanming Yuan*.

The 12th day of the 3rd month of the 61st year of Kangxi's reign (April 27, 1722) marked an important occasion in the history of Yuanming Yuan. The 11-year-old Hongli (i.e., Emperor Qianlong) met with his 68-year-old grandfather Emperor Kangxi at the Peony Terrace. As the boy was particularly favored by the old emperor, he was specifically allowed to accompany his grandfather and live in the palace. This means that Hongli had resided in Changchun Yuan for a short time. No one had expected that the emperor was not going to live long. Over the 8 months from the 3rd to the 11th months of the year, Hongli stayed beside his grandfather, not only when the emperor was handling state affairs in Changchun Yuan and the Forbidden City, but also when he was hunting at the South Palace and in places beyond the Great Wall. The short but beautiful moments spent together left a dear memory between grandfather and grandson.

❹ Nanmu (Phoebe zhennan) is a rare tree species found only in China. Among the Qing imperial gardens, a nanmu hall is preserved at the Imperial Mountain Summer Resort and North Sea of the West Palace.

This much-told meeting took place at the Carving the Moon, Tailoring the Cloud (named Peony Terrace earlier), one of the subsequent "Forty Scenes of Yuanming Yuan" (Fig.3.17). Emperor Qianlong used this name as a metaphor for his grandfather and father's enlightenment and indoctrination to him. Amid the full-blown peony shrubs stood a uniquely-shaped, particolored nanmu❹ hall—the Memorial Hall, built in memory of the two late emperors.

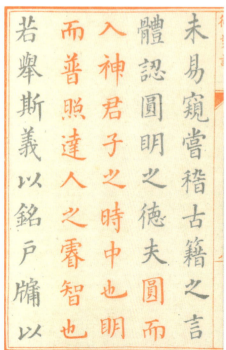

## The arrangement pattern of court at front, life at rear

In his poem *Justness and Honesty*, Emperor Qianlong made a high generalization of the essence of Yuanming Yuan: This "**place of interest**" is just like the garden of King Wen of Zhou. Actually, when Emperor Yongzheng expanded the garden, he inherited the philosophy adopted by Emperor Kangxi when building Changchun Yuan—This provides direct evidence for the association between the two gardens in the planning method.

The southernmost end of the entire garden was the core imperial court section. The place was intended for ceremonies, daily office work, and bringing up princes. The magnificent Grand Palace Gate, Secondary Palace Gate, and Hall of Justness & Honesty were disposed from south to north along the central axis. The scenic area "Diligence in Government, Affection to the Virtuous" was cored around the Hall of Diligence in State Affairs and located east of the Hall of Justness & Honesty. It was a place for the emperor to handle work, meet with his ministers and take a short break. Further east was the Depths of Fairy Caves, the place where the princes resided and studied (see more details in the Court at Front, Life at Rear section of Chapter 5).

Similar to Changchun Yuan, Yuanming Yuan also had its Front Lake and Rear Lake. The former functioned to separate the imperial living section from the court section with earth mounds and water surfaces. "The Nine Prefectures", which used to refer to nine regions in the Central Plains before, has become an alternate name for China. This concept was further reinforced by the *Books of Ancient Times-Tribute of Yu*, one of the earliest geographic books in China, and by the "Small Sea" and "Vast Seas" proposition of the famous Warring States scholar Zou Yan (c.324—250 B.C.). Now in Yuanming Yuan, there appeared a concretized downsized version of the Nine Prefectures, depicted on the soil ground. Nine varying-sized, diversely-shaped islands encircled the 200 m square Rear Lake. Streams ran both between and on the circumference of the nine islands. The total scenic area was about 190,000 m². That was the renowned Nine-Prefectures scenic area. Needless to say, the Rear Lake stood for the Small Sea while those streams and lakes out there stood for the Vast Sea. All these details were evidence for the scrupulous depiction of Emperor Yongzheng when building the garden.

On the largest-sized island in this scenic area, three halls were placed along the central axis of the court section—Hall of Yuanming Yuan (1# in Fig.3.18-a), Hall of Honoring Three Selflessnesses (2#) and Hall of Peace over the Nine Prefectures (3#). This arrangement was obviously inspired by the Forbidden City. On the west and east sides

**Fig. 3.15**
Calligraphic scripts of *Note of Yuanming Yuan* by Emperor Yongzheng

**Fig. 3.16**
Qing. Court Painter. *Scroll Portrait of Emperor Yongzheng Reading Books* (Palace Museum)

**Fig. 3.17**
Qing. Court Painter. *Carving the Moon, Tailoring the Cloud* in the *Album of Paintings and Poems of Forty-Scenes of Yuanming Yuan* (National Library of France)

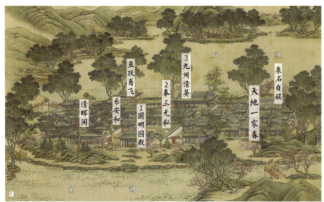

of the three halls spread the bedchamber buildings of the emperor and his concubines (Fig.3.18-a).

The other 8 islands in the Nine-Prefectures scenic area all had their particular themes: plants—Carving the Moon, Tailoring the Cloud (peony), Natural and Picturesque Scenery (bamboo), Academy of Phoenix Trees (phoenix tree), Winehouse of Apricot Blossoms in Spring (apricot flower, vegetables); worship—the Mercy Cloud of Universal Blessing; fish watching—the Magnanimousness of Mind; waterscape—the Merged Colour of Waters and Skies; books reading and collection—the Inclusiveness of Ancient and Modern.

The name "Natural and Picturesque Scenery" means that, against the background of the Western Hills, the entire Rear Lake scenic area looked like a naturally composed picture (Fig.3.18-b). The Strolling in Paintings at Qingyi Yuan also bore the same ideorealm. Among the buildings in these scenic areas, the three-story bell tower (1# in Fig.3.18-c) in the Mercy Cloud of Universal Blessing scenic area looked very special. At a south-facing point of the second story, there was a striking clock. It was probably the earliest European mechanical product in Yuanming Yuan. On top of the roof was a phoenix-shaped wind flag. With this flag, the emperor was able to tell the time and wind direction across the lake when he was in the Peace over the Nine Prefectures. It was indeed a very ingenious design (Fig.3.18-c). Besides, the lake-facing, two-story belvedere, together with the pavilion and curved bridges installed on water in the tower of the Merged Colour of Waters and Skies scenic area were also uniquely structured. Here one could not only look down from the tower, but also stroll over the waves (Fig.3.18-d).

**Fig. 3.18**

Qing. Court Painter. Part of the (a)Peace over the Nine Prefectures, (b)Natural and Picturesque Scenery,(c) Mercy Cloud of Universal Blessing and (d)Merged Colour of Waters and Skies in the *Album of Paintings and Poems of Forty-Scenes of Yuanming Yuan*

## Overall engineering philosophy of depicting natural landscapes

Besides scaling down the Nine-Prefectures, another unique originality of Yuanming Yuan was imitating the natural landscapes. In order to create an overall layout to simulate the geography of the empire—the terrain is higher in the northwest than in the southeast, the mountains originate from Mount Kunlun and the water flows southeastward into the sea—the only choice was to build artificial lakes and hills with manual labor. The tough reality was: the terrain of Yuanming Yuan was higher in the southwest and lower in the northeast.

First, topographically, the ancients created a commanding height (about 14 m above the ground level) at the northwestern corner of the garden with earth mounds and rockeries. They also laid out a Hill House of the Purple Jade Palace scenic spot, named after the Elixir Room of Purple Jade said to exist on Mount Kunlun. Next, all earth mounds in the garden originated from this place. They stretched and undulated either southward or eastward, despite interruptions by water surfaces. On their way, they produced two sub-heights at the northwestern corners of the Nine-Prefectures and Blessing Sea scenic areas.

Water in the garden came from two sources. The main stream of water came in from the southwest and flew northwest down the terrain. Then it redirected southeastward into individual scenic areas and finally poured into the large lake surface of the Blessing Sea in the east. The other stream of water was brought in from the guarding river at the northwestern corner of the garden. It formed a 1.5 km long serpentine diversion canal running across the garden from west to east. This stream of water was used to irrigate the farmlands opened in the garden and for local landscaping. It was similar to the farmland area at the west side of Changchun Yuan.

In order to simulate the ideorealm of the East Sea, Emperor Yongzheng ordered to dig a huge, 500 m square lake—the Blessing Sea. Its water area alone outsized the entire Nine-Prefectures scenic area. The name "Blessing Sea" may have either come from the legend of Xu Fu's eastward voyage in the Qin dynasty, or simply for the sake of good wishes ("fu" means blessings). To ensure the best sightseeing effect, he not only laid out scenic spots on the 11 large islands around the lake surface, at the center of the lake, but also reproduced the 3 legendary immortal islands on the East Sea—Penglai, Fangzhang and Yingzhou—by imitating the depiction of Tang dynasty painter Li Sixun (651–716). The place was named "Immortal's Residence on Penglai Island" (Fig.3.19). It was a typical "One-pond Three-mountains" matrix, constructed to express the monarchs' wishes for immortality and longevity since ancient times.

This huge water surface became a water activities center not found in the Forbidden City. Activities there included dragon boat races and floating river lanterns (Fig.3.20). As to land activities centers, apart from the Justness & Honesty and Peace over the Nine Prefectures, there were two other particular ones: Garden of Shared Happiness and the High Mountains & Long Rivers.

The name "Garden of Shared Happiness" conveys the emperor's wish to share happiness with his people. This recreational center of the imperial court consisted of a Marketing Street and a Grand Opera Tower. It was a place for daily catering, watching dramas and lanterns during festivities. The High Mountains & Long Rivers, another venue located in the southwest of Yuanming Yuan, often hosted activities requiring open spaces, such as fireworks, acrobatics and wrestling shows. It was one of the major venues for the emperor to inspect his armies, too (see more details in the Festival Celebrations section of Chapter 5). To highlight the ethnic policy of the empire, in the lake east of the High

Mountains & Long Rivers, Emperor Yongzheng designed an above-water building that looked like the Buddhist symbol "卍" in plane. It was perhaps one of the most uniquely-shaped ancient Chinese buildings. The name "Universal Peace and Harmony" was used to convey the Buddhist connotation and the wish for a peaceful world (Fig.3.21). The emperor himself enjoyed coming for a break, too.

Emperor Yongzheng was an earnest believer of Buddhism and Taoism. In addition to the Mercy Cloud of Universal Blessing, other religious buildings, including the Vairocana's Magnificent Residence, Sravasti City, and Prosperous Descendant Palace, were distributed at different points throughout the garden. Their sizes were much larger than those in Changchun Yuan. Obviously, the emperor wished that the country would be blessed by all gods and Buddhas (see more details in the Landscape Tours section of Chapter 5). Like his father, Emperor Yongzheng also embellished the large tracts of farmlands in the middle of Yuanming Yuan with scenic spots, including the Luxuriant Crops Pavilion, Crops Watching Pavilion, and Plowing and Weaving Pavilion, to show his emphasis on agriculture. A typical example was the horizontally checkered Simplicity and Tranquility in Mind (a 田-shaped house. "田" is the Chinese for "farmland"). It was indeed a very special building (Fig.3.22). In the north, he put up a North Distant Mountain Village in imitation of the water villages in regions south of the Yangtze River. He himself would often come and see how well the farm crops were growing, too.

After Emperor Qianlong was enthroned, he highly cherished this family property left by his father. He also incorporated his own garden-making ideas into Yuanming Yuan. In addition to remodeling a couple of scenic spots like the Hill House of the Purple Jade Palace and Broad Mind & Universal Justice, he also organized the construction of a new imperial ancestral shrine—the Peace and Blessing Palace (Great Mercy and Eternal Blessing), buildings for reading and collecting books—the Academy of Gathering Talents and Pavilion of Literary Fountainhead, as well as the Taoist Wonderland on Fanghu Island and Mirror-like Lake Reflecting the Sky in the Blessing Sea scenic area. The garden was made more functional and better arranged. The emperor also managed to unify all the names of the scenes with a four-character phrase.

The Peace and Blessing Palace was another large imperial ancestral shrine besides

**Fig. 3.19**
Qing. Court Painter. The Immortal's Residence on Penglai Island in the *Album of Paintings and Poems of Forty-Scenes of Yuanming Yuan* versus current status

**Fig. 3.20**
Qing. Court Painter. Boats sailing on the Blessing Sea depicted in the *Scroll Paintings of Forbidden Palaces in Twelve Months* (Palace Museum in Taipei)

**Fig. 3.21**
Qing. Court Painter. Universal Peace and Harmony in the *Album of Paintings and Poems of Forty-Scenes of Yuanming Yuan*

**Fig. 3.22**
Qing. Court Painter. Simplicity and Tranquility in Mind in the *Album of Paintings and Poems of Forty-Scenes of Yuanming Yuan*

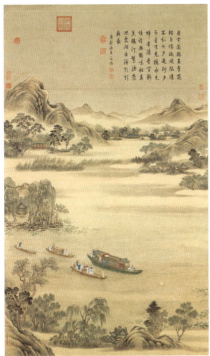
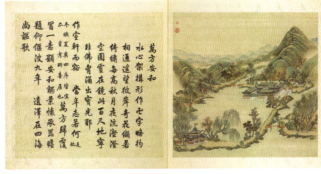
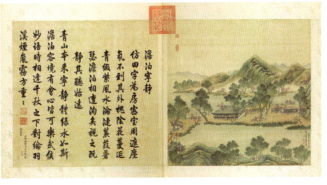

the Coal Hill inside Beijing City. It was specifically built to hang portraits of the late emperors and their empresses and concubines that had passed away. This was also a venue for holding worship ceremonies on the Lantern Festival, the birthdays, and the death anniversaries of those enshrined. This way, the northwestern corner of Yuanming Yuan was made into a large religious worship zone.

In the 9th year of Qianlong's reign (1744), the most important set of pictures of Yuanming Yuan—*Album of Paintings and Poems of Forty-Scenes of Yuanming Yuan*—was formally completed. The incomparably exquisite fine brushwork co-created by Well-content Mansion (Ruyi Guan) painters Tang Dai and Shen Yuan gave a live depiction of how the 40 scenic areas looked in different seasons. The pictures are also accompanied by poems of Emperor Qianlong transcribed by minister Wang Youdun, as well as garden notes written by Emperors Yongzheng and Qianlong. This art piece carries a big weight both in historical and artistic values. It was sacked to France during the 1860 warfare and is now kept at the National Library of France. Besides, the painting also derived a variety of other versions such as watercolor and woodblock paintings. Some derivatives were passed to Europe, spreading the name of Yuanming Yuan and its attractive scenes throughout the world (Fig.3.23).

**Fig. 3.23**
Qing, Chinese Court Painter and French missionary.
The traditional, watercolor and woodcut versions of the scene of Taoist Wonderland on Fanghu Island in Yuanming Yuan

In his *Postscript on Yuanming Yuan*, Emperor Qianlong complimented, "**Yuanming Yuan possesses all the best factors under heaven. It is second to none. There is no possibility of finding a better garden**"(Fig.3.24). Complacent as it may sound, this compliment is not in the least overstating. Right because of this fact, the emperor warned his offspring not to spend a lot of money and labor building imperial gardens anymore. Ironically, what he said did not mean to put an end to the garden making by the Qing imperial family. Instead, his interest in garden making had but just started. As

Fig. 3.24
Qing, Giuseppe Castiglione. Emperor Qianlong in the *Portrait of Emperor Qianlong and His Empress and Concubines* (Cleveland Museum of Art, US)

he was to build many other gardens in the years to come, this commitment turned out to be nothing but empty talk. After him, however, before Yuanming Yuan was ruined, his successors Emperors Jiaqing, Daoguang and Xianfeng did refrain from launching massive constructions except some local modifications. Even so, the money spent on these modifications, plus the daily maintenance of the garden, was also a large sum of expenditure.

... As for the pleasure-houses, they are really charming. They stand in a vast compass of ground. They have raised hills, from 20 to 60 foot high, which form a great number of little valleys between them. The bottoms of these valleys are watered with clear streams which run on till they join together and form larger pieces of water and lakes. They pass these streams, lakes, and rivers in beautiful and magnificent boats. I have seen one, in particular, 78 foot long and 24 foot abroad, with a very handsome house raised upon it. In each of these valleys, there are houses about the banks of the water, very well disposed with their different courts, open and close porticos, parterres, gardens, and cascades, which, when viewed all together, have an admirable effect upon the eye.

They go from one of the valleys to another, not by formal straight walks as in Europe, but by various turnings and windings, adorned on the sides with little pavilions and charming grottoes, and each of these valleys is diversified from all the rest, both by their manner of laying out the ground and in the structure and disposition of its buildings.

All the risings and hills are sprinkled with trees, and particularly with flowering trees, which are here very common. The sides of the canals or lesser streams are not faced, (as they are with us) with smooth stone and in a straight line, but look rude and rustic with different pieces of rock, some of which jut out and others recede inwards, and are placed with so much art that you would take it to be the work of nature. In some parts the water is wide, in others narrow; here it serpentines, and there spreads away, as if it was really pushed off by the hills and rocks.

The banks are sprinkled with flowers which rise up even through the hollows in the rock-work, as if they had been produced there naturally. They have a great variety of them for every season of the year. Beyond these streams there are always walks, or rather paths, paved with small stones, which lead from one valley to another. These paths too are irregular, and sometimes wind along the banks of the water, and at others run out wide from them.

On your entrance into each valley, you see its buildings before you. All the front is a colonnade, with windows between the pillars. The wood-work is gilded, painted, and varnished. The roofs too are covered with varnished tiles of different colors: red, yellow, blue, green, and purple, which by their proper mixtures and their manner of placing them form an agreeable variety of compartments and designs. Almost all these buildings are only one story high; and their floors are raised 2 to 8 foot above the ground. You go up to them, not by regular stone steps, but by a rough sort of rock-work formed as if there had been so many steps produced there by nature.

The inside of the apartments answers perfectly to their magnificence outside. Beside their being very well disposed, the furniture and ornaments are very rich and of an exquisite taste. In the courtyards and passages, you see vases of brass, porcelain, and marble, filled with flowers; and before some of these houses, instead of naked statues, they have several of their hieroglyphical figures of animals and urns with perfumes burning in them placed upon pedestals of marble.

Every valley, as I told you before, has its pleasure-house: small indeed, in respect to the whole enclosure, but yet large enough to be capable of receiving the greatest nobleman in Europe, with all his retinue. Several of these houses are built of cedar, which they bring with great expense, at the distance of 1500 miles from this place. And now how many of these palaces do you think there may be, in all the valleys of the enclosure? There are above 200 of them, without reckoning as many other houses for the eunuchs, for they are the persons who have the care of each palace, and their houses are always just by them, generally, at no more than 5 or 6 foot distance. These houses of the eunuchs are very plain, and for that reason are always concealed, either by some projection of the walls or by the interposition of their artificial hills.

All the canals are brought on to supply several larger pieces of water and lakes. One of these lakes is very near five miles round, and they call it a "meer", or sea. This is one of the most beautiful parts in the whole pleasure ground. On the banks are several pieces of building, separated from each other by the rivulets and artificial hills above-mentioned.

But what is the most charming thing of all is an island or rock in the middle of this sea, raised, in a natural and rustic manner, about 6 foot above the surface of the water. On this rock there is a little palace, which however contains a hundred different apartments. It has four fronts and is built with inexpressible beauty and taste; the sight of it strikes

# VII. Yuanming Yuan in the eyes of a French missionary who worked for the emperor

one with admiration. From it you have a view of all the palaces, scattered at proper distance round the shores of this sea; all the hills, that terminate about it; all the rivulets, which tend thither, either to discharge their waters into it, or to receive them from it; all the bridges, either at the mouths or ends of these rivulets, all the pavilions and triumphal arches that adorn any of these bridges, and all the groves that are planted to separate and screen the different palaces and to prevent the inhabitants of them being overlooked by one another.

The banks of this charming water are infinitely varied: there are no two parts of its alike. Here you see keys of smooth stone, with porticoes, walks, and paths running down to them from the palaces that surround the lake. There, others of rock-work that fall into steps contrived with the greatest art that can be conceived; here, natural terraces with winding steps at each end, to go up to the palaces that are built upon them; and above these, other terraces and other palaces that rise higher and higher and form a sort of amphitheatre. There again a grove of flowering trees presents itself to your eye, and a little farther, you see a spread of wild forest trees, and such as grow only on the most barren mountains. Then, perhaps, vast timber trees with their under-wood; then, trees from all foreign countries; and then, some all blooming with flowers, and others all laden with fruits of different kinds.

There are also on the banks of this lake, a great number of network houses and pavilions; half on the land and half running into the lake, for all sorts of water fowl. As farther on upon the shore, you meet frequently with menageries for different sorts of creatures; and even little parks for the chase. But of all this sort of things, the Chinese are most particularly fond of a kind of fish, the greater part of which are of a color as brilliant as gold; others, of a silver color; and others of different shades of red, green, blue, purple, and black: and some, of all sorts of colors mix together. There are several reservoirs for these fish, in all parts of the garden, but the most considerable of them all is at this lake. It takes up a very large space, and is all surrounded with a lattice work of brass wire, in which the openings are so very fine and small as to prevent the fish from wandering into the main waters.

To let you see the beauty of this charming spot in its greatest perfection, I should wish to have you transported here when the lake is all covered with boats; either gilt, or varnished, as it is sometimes, for taking the air; sometimes for fishing, and sometimes for jousts and combats, and other diversions, upon the water; but above all, on some fine night, when the fireworks are played off there, at which time they have illuminations in all the palaces, all the boats and almost on every tree. The Chinese exceed us extremely in their fireworks, and I have never seen anything of that kind, either in France or Italy, that can bear any comparison with theirs.

The above text is cited from a letter from French missionary Jean Denis Attiret (1702—1768) to his friend in Europe (Letter to M. d'Assaut in Paris), which gives a vivid view of Yuanming Yuan (Old Summer Palace) where he worked from a special perspective of a foreigner. He was obviously astonished and wild about how heavenly beautiful a classic Chinese garden could be.

# 3.4 Charngchun Yuan and European Palaces

**In this section, you will get an idea about:**
1. Why was the design idea of Charngchun Yuan very novel?
2. How did the European Palaces landscape integrate the European and Chinese garden styles?

## The abode of the septuagenarian Emperor—Charngchun Yuan

As the first accessory garden to Yuanming Yuan, Charngchun Yuan (The Hermit Eternal Spring's Garden) was a retirement garden built by Emperor Qianlong for himself immediately to the east of Yuanming Yuan. So, this garden named after his assumed name "Charngchun" was precisely his proprietary space. It functioned similarly as the Palace of Tranquil Longevity in the northeast of the Forbidden City.

By the 12th year of Qianlong's reign (1747), the landscaping layout and principal scenes throughout the garden had basically been completed. After that, he added 15 separate Chinese-style garden scenic areas to the existing landscape layout in succession. These included thematic scenic areas imitating private gardens typical of Yangzhou, Nanjing, Suzhou and Hangzhou (see more details in Special Topic VI). The entire garden was not fully completed until the 37th year of Qianlong's reign (1772), when the construction of the Lion Grove Garden, a scenic area having the same name as that in Suzhou, drew to a close above.

The entire Charngchun Yuan was almost square in shape. Its land area was about 760,000 m² (including the European Palaces scenic area), which was merely a third of the size of Yuanming Yuan. Its overall style was different from both the building compounds in the Palace of Tranquil Longevity and the fairly compact scenic spots in Yuanming Yuan (Fig.3.25). Here, water covered the absolute majority of the area. From the central island outstretched flexibly shaped combinations of causeways and islands. They produced lake surfaces of varying sizes and a number of scenic areas. The largest two square lake surfaces measured 60,000 to 70,000 m². They were larger than the Rear Lake at Yuanming Yuan, but far smaller than the Blessing Sea. On land, the large lake was encircled by continuous, undulated earth mounds on all sides. Several temples and garden-in-gardens were scattered in these earth mounds. Obviously, the design composition of Charngchun Yuan was very novel.

Restored plan of Charngchun Yuan (Xianfeng period) Fig.3.25

① Palace Gate
② Stoicism Hall
③ Classic Books Hall
④ Mansion of Exquisitely Jade-like Rock
⑤ Studio of Reflection in Clear Water
⑥ Garden of Taoist Fairyland
⑦ Studio of Considering Sustainability
⑧ Immortal's Mountain in the Sea & Broad Mind
⑨ Islet of Diffusing Fragrance
⑩ Temple of Dharma's Wisdom
⑪ Temple of Buddha's Solemn Appearance
⑫ Waterside Orchids Hall
⑬ Hall of Turning Boat Sail
⑭ Clustered Blossom Gazebo
⑮ Lion Grove Garden
⑯ Garden of Mirror-like Water
⑰ Imitated Zhan Garden
⑱ Garden of Flourishing Woods

Scenes of European Palaces
1  Perspective Bridge
2  Harmonious Wonder and Interest
3  Water Storage Tower
4  Maze of Engraved Brick Wall
5  Rare Birds Aviary
6  Landscape of Fairyland
7  Five Bamboo Pavilions
8  Hall of National Peace
9  Throne of Appreciating the Fountains
10 Magnificent Fountains
11 Distant and Oversea Landscapes
12 Stone Archway
13 Perspective Hill
14 Whorled Archway
15 Perspective Paintings on the Walls

Legend

Manmade earthen hill
Water
Building
Square
Contour line
Wall

Following the usual practice, the Palace Gate area of Charngchun Yuan was sited at the southernmost end. Compared with Yuanming Yuan, this place was much simpler both in layout and function. The main hall, the Stoicism Hall, sat on a platform. Behind the hall was the water-facing Pavilion of Shared Happiness, an opposite scene to the water pavilion on the other side of the lake. Emperor Qianlong placed his retirement bedchamber on the largest island at the center. This regularly-shaped giant group of palace buildings covered 18,000 m² of land. It was encircled by earth mounds on all sides. Interestingly, its north-to-south central axis was 100 m west of the axis of the Palace Gate. The layout was very flexible. Upon entry into the plaza from the eastern, western and southern archways, and passing through the palace gate, one would see the three halls of the Classic Books Hall, Chunhua Pavilion and Containing Trueness Studio, and the side halls to their east and west. One particular thing was that "Chunhua" was a reign title of the Song dynasty. The magnum opus in the ancient Chinese calligraphic history—the *Calligraphy Copybooks of Chunhuage*— was created at that time. Emperor Qianlong associated his palace hall with the calligraphic model, just in order to keep the inscribed calligraphic pieces from the past dynasties in the left and right ambulatories in front of the hall. It functioned similarly as the Hall of Happiness & Longevity at the Palace of Tranquil Longevity.

On the causeway islets surrounding the island, the groups of scenes each had their own features. The Immortal's Mountain in the Sea & Broad Mind was located on the circular island at the center of the lake. Immortal's Mountain means a fairy mountain on the sea. Hence it also carries the wish for immortality and longevity. Climbing up to this tall pavilion and looking west, one would get the full view of Yuanming Yuan and the beautiful scenery of the Western Hills. The Mansion of Exquisitely Jade-like Rock was situated on the mid-lake island east of the Classic Books Hall. Exquisitely Jade-like Rock was used to describe the finely shaped rockery. The Tranquil Cranes Studio and Taming Gulls Pavilion on the island both convey a leisurely mood of spending time with animals. The Grass-sized Boat was actually a miniature version of a painted pleasure boat house. Its name comes from a statement in *Zhuang-zi*. To the south, the Studio of Reflection in Clear Water was an irregularly-shaped building facing waters on three sides. Its waterfront arc-shaped corridor was the most stylish. Here one could acquire the best water-watching experience.

In the north of Charngchun Yuan, minor scenes such as the Temple of Dharma's Wisdom, Temple of Buddha's Solemn Appearance, Waterside Orchids Hall and Lion Grove Garden were arranged in a linear formation. These scenic spots were either hidden in the valley, or looking down from the height, or placed with waters in front and hills hind. They offered rich landscapes to this 700 m long artificial hill. On the other side, the Studio of Considering Sustainability, Garden of Flourishing Woods, Imitated Zhan Garden, and Garden of Mirror-like Water were laid out along the east-to-west trending, approximately 70 m wide watercourse. Touring the garden by boat here was very similar to what one can experience on the Slender West Lake of Yangzhou. Upon arrival at the wharf, one could enter the garden. There a completely different world would be waiting ahead.

## A hybrid of Chinese and European gardens—the European Palaces

On the strip in the north of Charngchun Yuan, there appeared something never seen before in the Chinese garden history—the European Palaces scenic area. Its total area was approximately 86,000 m². From the design of the Harmonious Wonder & Interest, the first European-style palace compound, in the 12th year of Qianlong's reign (1747) to the completion of the Distant & Oversea Landscapes, the last scenic spot there, in the 48th year of Qianlong's reign (1783), the construction of the European Palaces spanned 36 years. This architectural work was jointly built by European missionaries and Chinese craftsmen under the direct plotting and composition of Emperor Qianlong. It contained

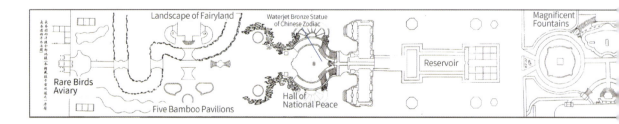

rich Chinese traditions and the cultural elements of the Renaissance. In a way, it was not only the product of the heydays of the Qing empire. It was also a representative outcome of exchange between eastern and western cultures.

European science and technology was imported into the Chinese imperial court along with missionaries as early back as in the early 17th century (Ming dynasty). In Kangxi's period, the emperor himself studied western sciences such as astronomy, mathematics, and anatomy. In Yongzheng period, western decorations and mechanical products began to serve the imperial gardens. The striking clock tower, the artificial waterfalls (Graceful Scenery of the Western Peaks), the water-driven fans (Clear Water and Flourishing Woods), and the various striking clock ornaments displayed indoors at Yuanming Yuan, were all typical representatives. Yet these did not make much difference to the overall style of the gardens. Hence the European Palaces at Charngchun Yuan, as a separate scenic area, was extraordinarily significant.

The entire scenic area was arranged in the east-to-west direction along the axis. The total length was 850 m. The distance between the northern and southern walls was only 60 m, making the area look quite slender. At the western end (the Harmonious Wonder and Interest, Fountain Pond, and Maze of Engraved Brick Wall) and the midmost point of this strip (Distant & Oversea Landscapes, Magnificent Fountains and Throne of Appreciating the Fountains), the space was enlarged. A north-to-south axis was formed to effectively relieve the feeling of depression. In order to avoid potential incompatibility between Chinese and European gardens, the scenic area was fully enclosed by tall bounding walls. Its southern side was located immediately adjacent to the earth mound north of Charngchun Yuan. This way, access was possible only from the Perspective Bridge at the western end and the Waterside Orchids Hall of Charngchun Yuan.

Despite its overall appearance as a European-style architectural complex, the European Palaces was actually a clever mix of both Chinese and European gardening approaches. The main tower of the Harmonious Wonder & Interest was enormously sized. It consisted of a three-story tower and two octagonal pavilions connected by corridors. In the begonia-shaped water pond at the front stood brass water-spraying animal statues, but the tower itself was flanked by traditional earth mounds and rockery arrangements (Fig.3.27). Before entering the Maze of Engraved Brick Wall, two stone lions were looking at each other at the European-style garden gate. The Maze of Engraved Brick Wall was actually a 90 m by 60 m rectangular labyrinth. It consisted of 卍-patterned parapets instead of hedge grows commonly found in Europe. The labyrinth was surrounded by a moat-like watercourse on three sides. At its northern side rose an earth mound with a Chinese-style pavilion on top. This was similar to the ideal geomantic matrix of ancient Chinese cities.

After walking through the Rare Birds Aviary, one would formally enter the slender area of the European Palaces(Fig.3.26). Emperor Qianlong named a two-story tower

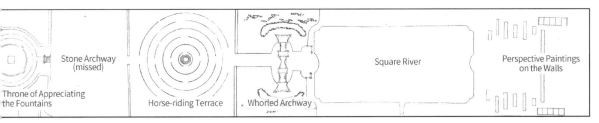

**Fig. 3.26** Qing, Style House. Partial Plan of the European Palaces at Charngchun Yuan

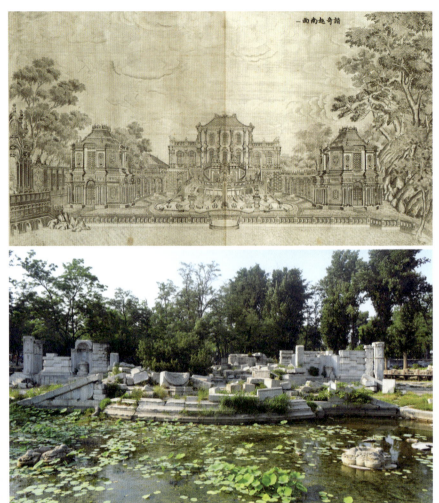

**Fig. 3.27**
Qing. Court Painter. The South Side of the Harmonious Wonder and Interest in the *Copperplate Etching of European Palaces* versus current status

Landscape of Fairyland, meaning it was a landscape beyond the noisy world and a location to live away from the earthly world. Here was the place where the emperor's Uyghur concubine Rongfei did her religious services. Not far away, the Hall of National Peace was a tall European-style palace (Fig.3.28). The name means a time of national peace and order, just similar to the name "Peace over the Nine Prefectures" at Yuanming Yuan. In order to highlight the theme of "sea", the architect smartly implanted the Chinese culture into the building. The top end of the inner side of the water pond was terminated by two fish with their tails entangled together. Water from their mouths fell down the edge of the pond. Below these fish was a huge scallop-like stone sculpture. Out of the mouth of this sculpture, water flew and fell down to the rockeries below. This was a symbol for the Haishuijiangya—a graphic pattern of Mountain in the Sea (common in Chinese imperial robes and official uniforms). More creatively, the pond was lined with 12 animal-faced, human-bodied brass statues, representing the 12 animal signs of Chinese Zodiac (Fig.3.29). From the detail, although they were carved more like European-style animal images, the statues were dressed and posed in the Chinese manner. The rabbit was waving a folding fan, the dog was grasping a bow and arrow, the snake was bowing down with its hands folded in front respectfully, the

Fig. 3.28
Qing. Court Painter. The Sest Side of the Hall of National Peace in the *Copperplate Etching of European Palaces* versus current status

monkey was holding its golden cudgel (symbolizing Sun Wukong). The fountain sprayed water automatically toward the center through different brass statues at different times of the day.

To the east of the Hall of National Peace sat the most renowned Magnificent Fountains (Fig.3.30). Here, the fountain was combined with a European-style white marble archway, with 11 symmetrically-arranged layers of brass water-spraying towers. The lion at the center of the archway opening sprayed water down, producing a cascade of beaded curtains. In the bat-like water pond at the front, a live hounds chasing a deer story from ancient Greek myths was being performed. Not only were all the 10 hounds spraying water toward the deer at the center, the antlers of the deer itself were also spraying water out. Although this theme originated from the western world, it coincided with the nomadic horse-riding and archery culture of the Manchu nationality. Not far away across from this fountain lay the platform for seating the emperor's throne and a floor screen. This platform did not follow the stereotyped old customs. Instead, it was placed the other way round to face the north. The floor screen was inlaid with five marble reliefs. On these reliefs, armors, cannons and swords were embossed as if they were showing off the military power

**Fig. 3.29**
The 7 existing water spray brass sculptures of the Chinese Zodiac at the Hall of National Peace (National Museum of China and Poly Art Museum)

of the empire. On top of the elevated stand behind the Magnificent Fountains rose a tall building named Distant & Oversea Landscapes. It means a landscape far beyond the seas. The design of the building blended a European-style bell tower and facade decoration with a Chinese-style large roof. The door was guarded by a pair of stone lions. The columns were carved with many decorative plant patterns. Emperor Jiaqing often toured here and wrote poems. To him, the European Palaces scene was the symbol of European countries acknowledging allegiance to the Chinese imperial court.

When a British mission came on a visit in the 58th year of Qianlong's reign (1793), the emperor made a specific arrangement for the mission to pay a visit with reverence to his European-style garden, just to show the stateliness of his "celestial empire".

During the same period in Europe, imperial families and aristocrats took pride in

Fig. 3.30
Qing. Court Painter. The South Side of the Magnificent Fountains in the *Copperplate Etching of European Palaces* versus *current status*

possessing Chinese-style utensils or ornaments. They even tried to build Chinese-style garden landscapes and buildings in their own gardens. Today we can still see the Great Pagoda in Kew Garden, designed by the eclectic British architect William Chambers (1723—1796) who visited China. The "East learning from the West" on the part of China and the "West learning from the East" on the part of Europe marked a wonderful episode in world's cultural history. Yet this did not make the European Palaces luckier to escape the gunfire of the European invaders.

Here in Charngchun Yuan, Emperor Qianlong planned himself an unusually wonderful life as a retired emperor. On the 1st day of the 1st month of the 1st year of Jiaqing's reign (1796), Emperor Qianlong declared to the world his decision to relinquish the throne. However, he did not really passed his power to his son. The new emperor could do nothing but remain at the beck and call of his father. Three years later (1799), on the 3rd day of the 1st month, the 89-year-old Emperor Qianlong passed away in the Hall of Mental Cultivation of the Forbidden City after ruling China for 63 years. Left behind was the enormous family property of the THFG and a perilous Qing empire. The history of the Qing dynasty also turned to a new page.

# 3.5  The darkest days

**In this section, you will get an idea about:**
1. For what purpose did the Anglo-French Allied Forces launch a war of aggression against China?
2. What kind of a catastrophe did the THFG suffer in this war?

The THFG was the highest artistic achievement in ancient China created and managed through painstaking elaboration by the Qing empire for more than 200 years. Tragically, it was destroyed almost in its entirety in the war of aggression against China launched by Britain and France in the 10th year of Xianfeng's reign (1860). This cultural catastrophe engraved an unforgettable memory in the history of the Chinese nation.

## The 1860 war of shame

To trace the source of this war, we have to look back to the 58th year of Qianlong's reign (1793) and the 21st year of Jiaqing's reign (1816), when Britain sent two missions to China. On the former mission, British embassy George Macartney showed to Emperor Qianlong the then state-of-the-art technological products in Yuanming Yuan, including carriages, cannons and astronomical instruments, hoping to open the door of the Chinese market. But the Qing empire was satisfied that it was a "celestial empire" and refused to regard Britain as a country of equal status. They had no interest in these products, either (instead, the European timepieces and other artwares were very appealing to the imperial family). Hence, they rejected the request for establishing trade ties. On the other side, ceramics, silk, and tea produced in China were sold to Europe in large volumes. Over time, there was already a huge trade gap between China and Britain. In Jiaqing period, some illegal British tradesmen began smuggling opium. They expected to reap fabulous profits by this despicable means. Before long, opium was already appearing in every part of the country, devouring the health of millions of Chinese people. A turning point appeared in the 19th year of Daoguang's reign (1839), when the Qing government destroyed 19,176 boxes plus 2,119 bags of opium illegally smuggled from Britain. The second year, the British government brazenly launched the First Opium War, which ended with China's failure and loss of land for reparation.

More than a decade later, the Second Opium War broke out. After failing in their

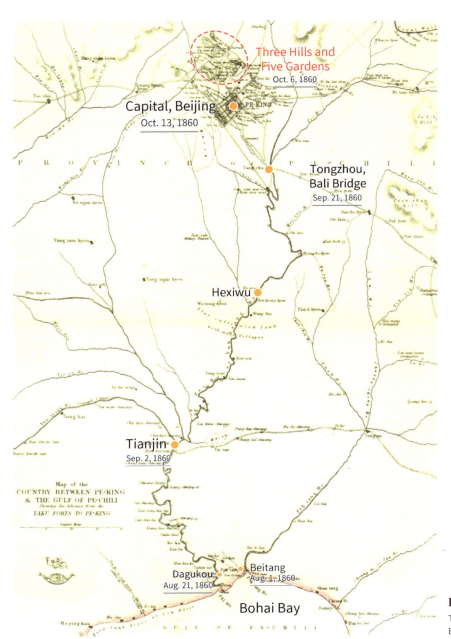

Fig. 3.31
The British battle map of Beijing and Tianjin (1863)

attempt to exchange treaties with the emperor in Beijing, Britain and France sent 13 warships into the White River of Tianjin. There they were greeted by heavy resistance from the Qing army.

In 1860, the ill-fated Qing empire was once again shrouded in the shadow of war (Fig.3.31). Only a few months after their defeat, Britain and France launched this retaliative war of aggression, numbering about 11,000 men from Britain and 6,790 from France (Fig.3.32). The catastrophe to the THFG took place was the consequence of the previous complicated political, diplomatic and military background. Clearly, both the

Fig. 3.32
British commander James Hope Grant (1808—1875) and his French counterpart Charles Cousin Montauban (1796—1878)

"revenge" and "punishment", as claimed by the Anglo-French Allied Forces, are unjust. All their words and deeds served just to unmask their real faces.

In August, the Qing army squared off again for the Anglo-French naval force at the coastal Fort Dagukou near Tianjin with full confidence, but ended in a fiasco because of strategic failure. In September, the Anglo-French Allied Forces started off for Tianjin along the White River. After losing the war, Emperor Xianfeng decided to send an imperial envoy to negotiate in Tongzhou near Beijing. However, the negotiations broke down. The next day, the Anglo-French Allied Forces raided Zhangjiawan Town. The Qing army was defeated and made their final stand at the Bali Bridge. By then, the Qing army had captured the Anglo-French delegation of 39 persons and removed them to the western suburbs' Haidian. On Sep. 21, despite its great advantages in number of men and geography and its dumbfounding bravery, the Qing army was defeated in the Battle of the Bali Bridge due to fatal tactical and equipment shortfalls. The battle was unimaginably fierce, with casualty tolls contrasting vastly between the two sides. On Sep. 22, despite the oppositions of his ministers, Emperor Xianfeng, fled to the Imperial Mountain Summer Resort in Rehe (Chengde). under the escort of 2,000 imperial guards, by excuse of autumn hunting.

On Oct. 5, the Anglo-French Allied Forces arrived in the City of Beijing. There they learned from the captives that the Qing army had fled to Yuanming Yuan. This marked the prelude to the tragic fate of the THFG.

### The looting of Yuanming Yuan

On Oct. 6, the French and British troops arrived at Yuanming Yuan in succession. At that time, the gates of the garden were tightly shut. Nobody was sure how many armymen were hidden in there. But when they tried to enter across the wall, the troops were heavily resisted by some 20 eunuchs. The imperial guards were nowhere to be found. Very soon, this huge imperial museum of Yuanming Yuan fell in its entirety into the hands of the invaders. Wenfeng, chief minister of the Imperial Household Department who was in charge of Yuanming Yuan, threw himself into the Blessing Sea, giving his life for the country.

In the diaries of members of the Anglo-French Allied Forces [5], they used a lot of flattering words to describe how beautiful the garden was and how they amazed at the gardening art of China. Yet this does not mean that they would choose to cherish or protect the garden. Instead, in their eyes, everything there was but a "trophy" that they could justifiably loot and destroy.

As narrated by M'Ghee, in the beginning, "**everything in the garden was kept intact.**" But later on, General Grant tacitly allowed the accompanying officers to take whatever they liked as a souvenir. Everyone was readily prepared to take this privilege and they did everything to satisfy their burning desire. The robbery grew out of control. Although they knew that this would result in bad consequences, the two generals did not make up their mind to stop the soldiers from robbery. They simply tried to talk their men out of doing so symbolically. The Anglo-French armies also organized a committee of 6 members to select the most valuable gifts for their monarchs. This way, from the 7 to the 8 of October, the looting of Yuanming Yuan went on for two whole days.

More infuriatingly, Chinese robbers and paupers also joined in the looting of the gardens. Disasters not only fell upon Yuanming Yuan, the Longevity Hill cum Qingyi Yuan, Jade Spring Hills cum Jingming Yuan and the stores nearby were all sufferers of the looting. Before these bandits withdrew from Yuanming Yuan, almost everywhere that could hold something was filled up with booties. An appreciable part of the treasures were also rudely destroyed. In his diary, Irisson felt sorry for the garden and the treasures. Yet he sounded more like regretting that they had not packed the entire garden away.

[5] Citations here are mainly based on the book *How We Got to Pekin: A Narrative of the Campaign in China of 1860* by Robert John L. M'Ghee, a chaplain to the armed forces on the 1860 expedition, and the *Le Journal d'un interprète en Chine* by French interpreter Maurice Irisson (in the *Translations of Memories on the Sufferings of Yuanming Yuan* published by Shanghai Zhongxi Book Company).

## The burning of Yuanming Yuan and the Three Hills

As recalled by Irisson, 20 out of the 39 Anglo-French delegation members were not sent back alive. What happened to them? In Jixian Yuan to the south of Yuanming Yuan, the Qing officers and soldiers did beat and humiliate the British and French prisoners when interrogating and jailing them. The captives' belongings were also sent to Emperor Xianfeng and kept at the Hall of Justness & Honesty in Yuanming Yuan. Some soldiers participating in the looting happened to discover the fact. The British army, as an illegal invader and the winner, did not think about the tremendous damages to the lives and properties of the Chinese people caused by the war they launched. Instead, they called themselves "victims" and even took revenge by destroying the imperial gardens.

On Oct. 17, when negotiation fell in a deadlock, James Bruce (8th earl of Elgin), the highest-ranking diplomat and leader of the British mission to China, decided to burn the entire compound of Yuanming Yuan to the ground (his father was reputed for looting stone carvings of the Acropolis of Athene). He also tried to talk the French into joining in their action, but was rejected by Jean-Baptiste Louis Gros, the highest-ranking diplomat of the French mission to China. Even so, the French should still be condemned no less strongly, as they had burned some of the scenic spots during their earlier looting action. As analyzed by historians, as winter was on the way, in order to cause the treaties to be signed with the Qing government and withdraw from Beijing in the shortest time possible, burning the

**Fig. 3.33**
The burned-down Fragrant Hills Temple at Jingyi Yuan

**Fig. 3.34**
The surviving temples and pagodas at Jingming Yuan

**Fig. 3.35**
Felice Beato. The burned-down Longevity Hill (1860)

**Fig. 3.36**
Ernst Ohlmer. The burned-down Harmonious Wonder and Interest at the European Palaces (1873)

**Fig. 3.37**
The China Pavilion in the Fontainebleau Palace in France

gardens would mean a heavy spiritual blow to the Qing government.

On October 18, a large British column of about 3,500 men marched to the western suburbs of Beijing. They set on fire Yuanming Yuan, the Longevity Hill cum Qingyi Yuan, Jade Spring Hills cum Jingming Yuan, and Fragrant Hills cum Jingyi Yuan. Inconceivably, the victims of the war were nothing related to defense installations, but were the centuries-old gardens and the architectural masterpieces in them. Yet their builders and owners did not have any strength to protect them. What a miserable strategy! In the British authors' accounts, when they saw how the garden was being burned, they felt it was an offense of nature and a devastation of creature; many years later, they regretted having given a fair but too severe punishment. This hypocritical and even ridiculous statement is nothing more than trying to find some peace of mind.

"**Seeing this, a feeling of sadness and indignation filled the whole heart. Throughout the ages, the outstanding buildings loved by countless people have been destroyed and will never be seen again. These buildings show the skills and styles of the past. They are unparalleled in the world. You saw them once and you will never see them again. They have been reduced to ashes and can no longer be reproduced by man. You turn around and can't bear to see it again**" (*general idea of M'Ghee's diary*).

In the fire that lasted for two whole days, the overwhelming majority of the buildings in the 4 large imperial gardens were burned to ashes. The crimes of the invaders have been

clearly recorded in numerous old, black-and-white photographs (Fig.3.33-Fig.3.36). The purgatory-like miserable scenes are far beyond words. In this "1860 Event", the Chinese people were greatly shocked, especially the scholar-bureaucratic class at that time.

According to a statistic list by Imperial Household Department officials and old photographs recently discovered, among the 3 gardens of Yuanming Yuan, Charngchun Yuan and Qichun Yuan, only some 20 remotely-situated building complexes were spared by the arson. The great mass of the cultural relics and treasures in these gardens were also lost in the war. According to the official statistics of that time, apart from Yuanming Yuan, among the 87,781 furnishings in Qingyi Yuan, Jingming Yuan, Jingyi Yuan and the Temple of Azure Clouds, and so on, 75,692 were lost, 9,596 remained intact, and 2,493 were impaired.

The burning of Yuanming Yuan and the Three Hills did harvest the fruit desired by the invaders. Although Emperor Xianfeng was far away in Rehe, the heavy blow he felt was as near as if he were on the spot himself. He ordered to agree to all conditions of Britain and France as soon as possible. On Oct. 24 and 25, the two countries each signed *The Treaty of Beijing* with the Qing government.

The Anglo-French troops did not withdraw from Beijing until early November. On their way to Europe from Beijing, the treasures of the imperial gardens were once and again sold at cheap prices. Till today, they are still scattered in the hands of many museums and private collectors throughout the world (Fig.3.37).

## Museological institutions having collections of the THFG relics (incomplete statistics)

| China | Outside China |
|---|---|
| Palace Museum, Palace Museum in Taipei, National Museum of China, Museum of Chinese Garden and Landscape Architecture, National Library of China, Exhibition Hall of Yuanming Yuan, Poly Art Museum, Peking University, Shenyang Imperial Palace Museum, Baoding Zoo, etc. | British Museum, British Library; National Library of France, China Pavilion in the Fontainebleau Palace, Guimet Museum, Invalides in Paris (France); Museum of East Asian Art (Germany); Metropolitan Museum of Art, Boston Museum of Fine Art (U.S.); Osaka Municipal Museum of Art, Toyo Bunko (Japan), etc. |

Note
Recommended reading material: *Who Collects Yuanming Yuan* by Liu Yang

### Current plan of the ruins of Yuanming Yuan (2020) Fig.3.38

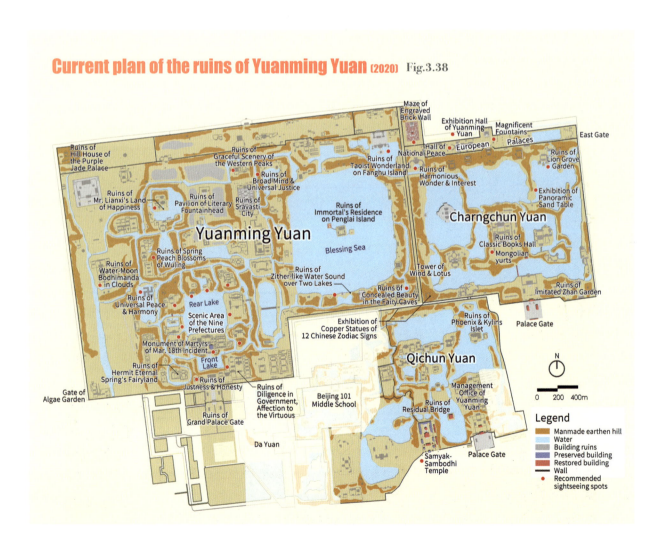

## International comments on the disgusting conduct

Famous Chinese historian Zhu Weizheng remarked, "**The burning of Yuanming Yuan was extremely cruel to China. The British and French colonial armies were extremely ignominious. Burning the garden was a destruction to civilization. The fire starters were civilized scoundrels.**" Yuanming Yuan scholar Zhang Chao noted, "**Brazen brigands shamelessly regard barbarity as a feat of greatness, while the righteous ones are furious about this act**". In his *Letter to Captain Butler* in 1861 regarding the Anglo-French Troops Expedition to China, French great writer Victor Hugo reprimanded the outrage of the Anglo-French Allied Forces:

"**One day two bandits entered the Old Summer Palace. One plundered, the other burned. Victory can be a thieving woman, or so it seems. The devastation of the Old Summer Palace was accomplished by the two victors acting jointly. Mixed up in all this is the name of Elgin, which inevitably calls to mind the Parthenon. What was done to the Parthenon was done to the Old Summer Palace, more thoroughly and better, so that nothing of it should be left. All the treasures of our cathedrals put together could not equal this formidable and splendid museum of the Orient. It contained not only masterpieces of art, but masses of jewelry. What a great exploit, what a windfall! One of the two victors filled his pockets; when the other saw this he filled his coffers. And back they came to Europe, arm in arm, laughing away. Such is the story of the two bandits.**"

## A brief attempt to rebuild Yuanming Yuan

In the 12th year of Tongzhi's reign (1873), the emperor attempted to rebuild Yuanming Yuan secretly by the excuse of using the garden as the adobe for the empress dowagers to enjoy their later years. He ordered the Imperial Household Department to make quick preparation for the design and construction. Some buildings had even taken initial form. However, as the budget was tight, the emperor not only called on his ministers to donate money, but also ordered to demolish the surviving building materials at the Three Hills and the granted gardens around Yuanming Yuan. At a time when the Qing empire was withering away, this was obviously an absurd and fruitless decision.

The effort to rebuild Yuanming Yuan declared a frustrating failure in less than a year because of exhausted budget and intensifying opposition. Not only was Yuanming Yuan unable to show its past brilliance, the other imperial gardens became victims, too. In the late years of the Qing dynasty, Empress Dowager Cixi had many a time toured Yuanming Yuan with Emperor Guangxu. Obviously, they still had a deep emotional attachment to the garden.

After the demise of the Qing dynasty, Yuanming Yuan was completely reduced to an unattended "construction material marketplace" and suffered decades of human destruction. Buildings, bounding walls, plants, rockeries, and anything else that could be reused were moved away by various people for various reasons. The hills and waters in the gardens, which were most representative of artistic values, were largely razed to the ground and used as farmlands or hamlets. That explains why the ruins of Yuanming Yuan today look so desolated and why the building components of the garden have made their way into many parts of the country (Fig.3.38).

# Touring the Three Hills and Five Gardens

Landscape, Art and Life of
Chinese Imperial Gardens

# Chapter 4
# From Qingyi Yuan to Yihe Yuan
—Garden Construction Originally Intended for Water Conservancy

# 4.1 Tracing back to the origin

**In this section, you will get an idea about:**
How was the water conservancy project established in the western suburbs of Beijing cored around the THFG?

Whenever we talk about the imperial gardens still existing in Beijing, apart from the Forbidden City and North Sea (Beihai), the next thing that comes up to our mind would be the well-known world cultural heritage—Yihe Yuan (Summer Palace)(Fig.4.1). In fact, this famous garden is a reconstruction during Guangxu's reign(1875—1908) from the ruins of the former Qingyi Yuan (Garden of Clear Water and Ripples), after Emperor Tongzhi failed in rebuilding Yuanming Yuan. As Yihe Yuan was built the latest in time, it is also the only well-preserved imperial garden among the THFG. Hence, Yihe Yuan represents the one of the most outstanding artistic achievement of the Chinese ancient imperial gardens and the remnant treasure of the THFG. Let's trace back to the two major historic events in Qianlong period, to look through the essence and origin of Yihe Yuan.

**Fig. 4.1**
Wu Xiaoping. Full view of the Longevity Hill after snow

In Yihe Yuan, there used to be a low-rise barren hill. Up there, according to legends, someone cut out a stone urn with "**flower and insect carvings**". For this reason, the hill was named "Urn Hill". A short distance away from the hill, there was a lake as old as least 2,300 years. As the lake lay west of the capital city, it was referred to as the West Lake or Urn Hill Lake. Prior to the 15th year of Qianlong's reign (1750), the names "Longevity Hill" and "Kunming Lake" were not existent. The following statements of Emperor Qianlong directly explain how names of the hill and lake came into being:

"**After the lake was completed, I named the hill and lake Longevity Hill and Kunming Lake. These names also carry my admiration for Emperor Yao and my wish to drill the navy, too.**"

"**In order to celebrate the 60th birthday of the Empress Dowager, I renamed the Urn Hill 'Longevity Hill'. I have also built a Temple of Immense Gratitude and Longevity on the southern side of the hill.**"

—— The *Note of Kunming Lake at the Longevity Hill* by Emperor Qianlong

## A birthday present

Temple building at the Longevity Hill started as early as in the Yuan dynasty. Up to the heydays of Qing, a magnificent array of temple buildings and garden scenes, collectively labeled as Qingyi Yuan, began to register their presence on the Longevity Hill. The very pioneer was the Temple of Immense Gratitude & Longevity seated on the south. This temple was built in celebration of the 60th birthday of the Empress Dowager in the 16th year of Qianlong' reign (1751). As his mother was a devout follower of Buddha, the Emperor, with "**endless gratitude**" to her, believed that, compared with simply holding birthday banquets or sending birthday presents, it would be much more meaningful to use this temple to express his appreciation of her kind parenting and his wish for her health and longevity (see more in 4.2 Qingyi Yuan).

## Launching water conservancy projects

In actual fact, however, the fundamental cause behind the building of Qingyi Yuan was water control. In Zhiyuan period of the Yuan dynasty, renowned scientist Guo Shoujing (1231—1316) made a marvelous contribution to the water conservancy engineering of the Great Capital of the Yuan dynasty (Fig.4.2-Fig.4.3). The Baifu Weir, a diversion canal excavated under his auspices, carried endless flows of spring water from the Baifu Spring in Changping to the semi-artificial reservoir at the Urn Hill Lake. From there, the water was carried further into the southeastern capital city through the Long River excavated during the Jin dynasty. It finally poured into the Grand Canal through the newly excavated Tonghui River to supply water for grain transport. Such a gigantic water conservancy system was almost the lifeblood of the Beijing City. It also underlay the rise of the scenic attractions in the northwestern suburbs. In the Ming dynasty, the Baifu Weir gradually fell into disuse. The spring water from the Jade Spring Hills kept the water conservancy hub of the Urn Hill Lake in operation (Fig.4.4). This place, owing to its beautiful scenery, once became a pleasure site for emperors and celebrities. In the book *Travel Jounal in Chang'an*, the scenery here was hailed as "**a spectacular sight comparable to the West Lake of Hangzhou**". Since the

**Fig. 4.2**
The ruins of the Baifu Spring in Changping

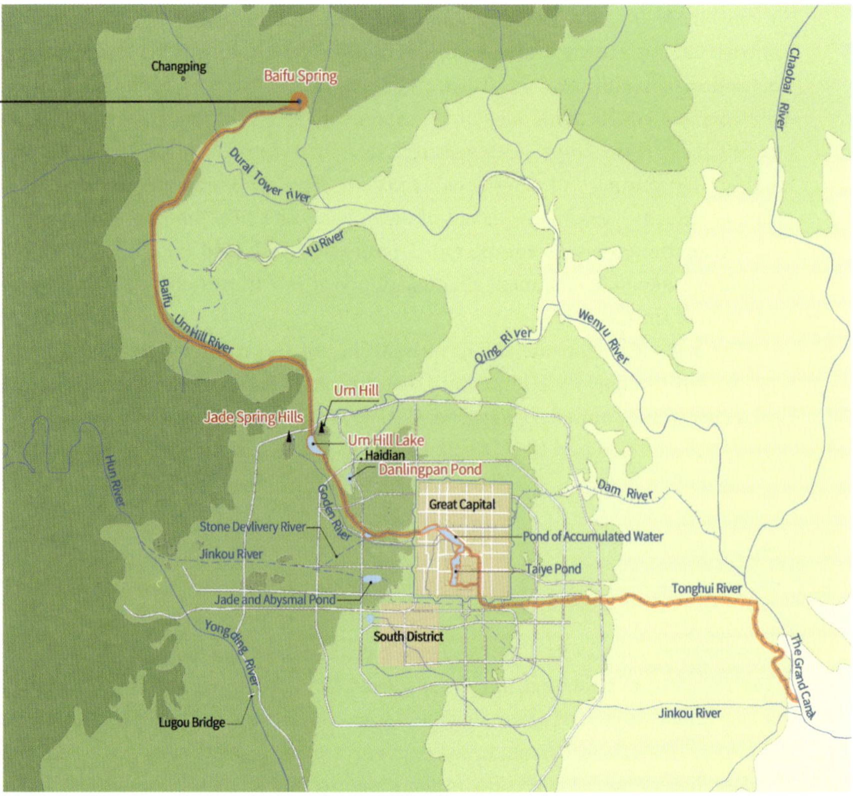

**Fig. 4.3**
Plan of the water conservancy system of Beijing in the Yuan dynasty (Redrawn from Beijing Historical Atlas by Hou Renzhi)

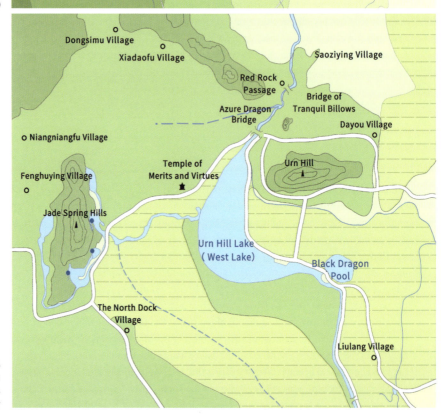

**Fig. 4.4**
Plan of the Urn Hill and Jade Spring Hills in the Ming dynasty

building of Changchun Yuan in the Qing dynasty, as the imperial family grew more and more interested in living in the environmentally comfortable western suburbs' Haidian, more and more smaller gardens had begun to sprout in the vicinity. Above all, the bulk-sized Yuanming Yuan in Yongzheng period rendered unprecedented strategic significance to this area.

In his article, Emperor Qianlong wrote, "**Rivers and canals are matters of national significance**". After a detailed investigation by the emperor himself, a grand water conservancy program to change the visage of the THFG came into being. Over the next ten years from the 14th year (1749) to the 24th year of Qianlong's reign (1759), the program was gradually brought to reality (Fig.4.5).

### Step 1: Lake digging and hill piling

Starting from the winter of the 14th year of Qianlong's reign (1749), farmers in their unworking months were called in by the imperial family. They took only 2 months to broaden the frozen Urn Hill Lake into a huge, 2,200,000 m² lake surface. The lake was shaped to imitate the "one causeway six bridges" layout of the West Lake in Hangzhou (Fig.4.6). A 2 km long West Causeway separated the lake into two parts. The water in the west overflowed into the larger part on the east side. This West Causeway lay right in the location of the original East Causeway of the Urn Hill Lake. The spring water from the Jade Spring Hills flowed eastward into the Kunming Lake by way of the Jade River and was stored in the lake. The water level of the Kunming Lake was jointly controlled by a number of sluices, including the sluice on the Azure Dragon Bridge at the northern end and the Two Dragons (Two-arch) Sluice on the East Causeway. When the sluice was opened, water could be discharged into the paddy fields or released during flood seasons.

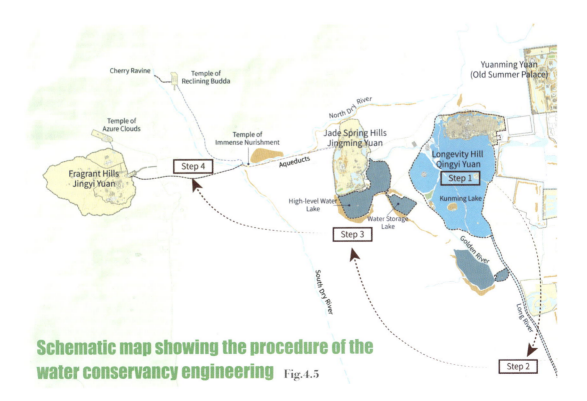

Schematic map showing the procedure of the water conservancy engineering Fig.4.5

The Kunming Lake is also believed by geographers as the "**first artificial reservoir in the suburbs of Beijing**".

The mud from lake excavation was also utilized to the fullest. After the artistic reformation, the unimpressive Urn Hill changed into a magnificent Longevity Hill in the blink of an eye. The reformation also laid good foundation for the subsequent landscaping of Qingyi Yuan.

### Step 2: Dredging up the Long River

Emperor Qianlong thought that increasing the water volume in the upper reaches could absolutely make water transportation possible between the capital city and Haidian. So, through 3 years of construction beginning from the 19th year of Qianlong's reign (1754), the 8.5 km long watercourse took on a completely new look. The river water ran past the Charngchun Bridge (Eternal Spring Bridge), Maizhuang Bridge (Wheat Village Bridge), Guangyuan Sluice (Extensive Source Sluice), Baishi Bridge (White Stone Bridge) before finally reaching the Gaoliang Bridge at outside Xizhi Gate of the capital. To provide pleasure functions, he ordered people to build a Yihong Tang (Hall of Leaning on Rainbow) palace and a wharf at the starting point of the Long River (Fig.4.7). He also rebuild the Longevity Temple which dated back to the Ming dynasty. This way, the Guangren Gong (Taoist Temple of Immense Benevolence), Changyun Gong (Taoist Temple of Prosperous National Fate ), and Longevity Temple, and so on could be seen halfway.

### Step 3: A second lake broadening

In order to solve the water shortage problem in spring, the emperor ordered people to open up two more lake surfaces: High-level Water Lake and Water Storage Lake. This project took 3 years and was completed in the 24th year of Qianlong's reign (1759). To prevent the High-level Water Lake from flooding, he used the Golden River as a drainage river and opened a lakelet at the southern end. The lake water poured into the Long River through the overflow dam at this point and flew further into the capital city.

### Step 4: Erecting aqueducts to divert water

In the 22nd year of Qianlong's reign (1757), the last massive water conservancy project was formally completed. Through the aqueducts, the spring water from the

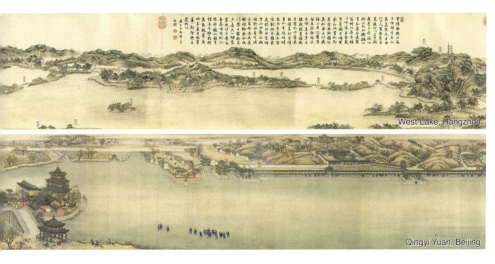

Fig. 4.6

Qing. Court Painter. The *Panorama of Ten-Scenes of the West Lake* (Palace Museum in Taipei) in comparison with the *A Panorama of Empress Dowager Chongqing in Birthday Celebration* (Palace Museum)

Fig. 4.7

Old photograph of Yihong Tang and the Long River

Fragrant Hills and Temple of Azure Clouds, and that from the Temple of Reclining Buddha converged into the pool at the Temple of Immense Nurishment. The water then flowed northeastward into Jingming Yuan. To ensure that the water would be well transmitted, the ancient people made smart arrangements. At high altitude, the aqueducts were placed on flat ground and coated with stone tile (possibly to keep it clean); at low altitude, they were placed on top of the wall (Fig.4.8-Fig.4.9).

Besides, near the Temple of Immense Nurishment, there were two dry rivers. One flowed northeast into the Qing River; the other flowed southeast into the location of the Jade & Abysmal Pond and finally poured into the moat. Although they were arid in normal times, these rivers were important drainer during flood seasons, safeguarding the capital city and imperial gardens downstream.

Through these 4 steps, the water conservancy engineering in the western suburbs was eventually completed. Although it cost a great amount of work and spanned a prolonged period of time, it provided a fundamental solution to the water conservancy problem of the Beijing City and the THFG (Fig.4.10). The entire water conservancy system can be rated as another milestone case in the science and technology history of ancient China after the great cause made by Guo Shoujing in the Yuan dynasty. More importantly, Qingyi Yuan was also basically completed between the 15th year and the 26th year of Qianlong's reign (1750—1761). Meanwhile, the expansion of Jingming Yuan also drew to a close in the 22nd year of Qianlong's reign (1757).

**Fig. 4.8**
Hou Renzhi. The ruins of the aqueducts at the Western Hills

**Fig. 4.9**
Old photograph of the paddy fields between the Jade Spring Hills and Longevity Hill (ROC period)

**Fig. 4.10**
Qing. Hongwu. Part of the *Panorama of Water Conservancy in the Capital* (National Museum of China)

1 Fragrant Rock Temple
2 Main Palace Gate
3 Tower of Reflection in Lake
4 Aqueducts
5 Glazed Pagoda
6 Pagoda of Lotus Treasury World
7 Dipamkara Pagoda
8 Sumeru Pagoda
9 Archway of Picturesque Scenery
10 Tower at Lake Boundary

# 4.2 Qingyi Yuan

**In this section, you will get an idea about:**
1. Why do we say the temples at the Longevity Hill made a "real-life" Buddhist world?
2. What were the respective features of the Front Hill, Rear Hill and lake area of Qingyi Yuan?

The importance of Qingyi Yuan lies in the following two aspects. First, how to deal with water conservancy engineering in a scientific, artistic manner so as to incorporate poetic illusions into the hills-and-waters landscapes; second, how to further combine it with political and military functions so that garden landscaping would serve Emperor Qianlong's feudal regime—Toward both points, Qingyi Yuan handed in an almost perfect "answer sheet" (Fig.4.11-Fig.4.12).

## Religious building complex

The Temple of Immense Gratitude & Longevity stands right in the middle of the Longevity Hill (Fig.4.13). It was undoubtedly the most important scenic spot in

11 Azure Dragon Bridge
12 Belverdere of Mirroring State Affairs
13 West Causeway
14 Hall of Appraising Talent
15 Phoenix Mound
16 South Lake Island
17 Seventeen-Arch Bridge
18 Spacious Pavilion
19 East Causeway
20 Tower of Wenchang Emperor

Restored plan of the Longevity Hill cum Qingyi Yuan (Xianfeng period) Fig. 4.11

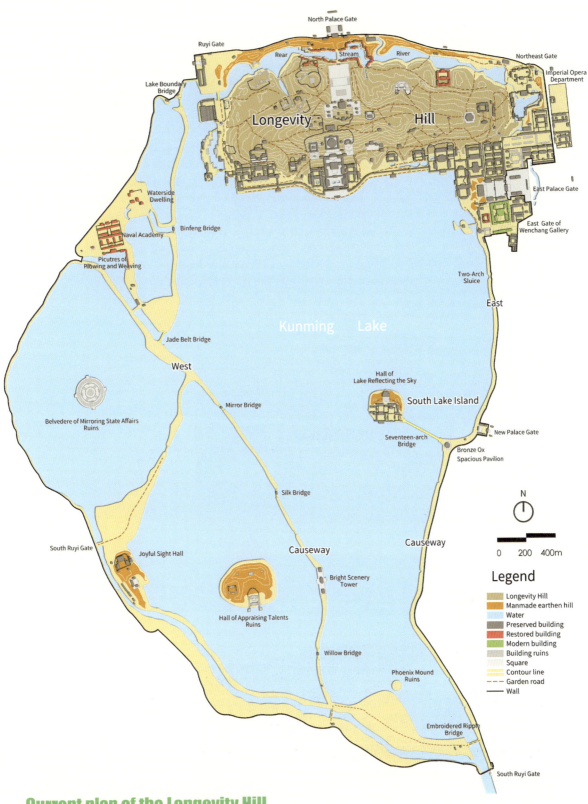

**Current plan of the Longevity Hill cum Yihe Yuan (Summer Palace)** (2020) Fig.4.12

(See the Current plan of the Longevity Hill in Fig.4.33)

**Fig. 4.13** Qing. Court Painter. The Temple of Immense Gratitude and Longevity in the *A Panorama of Empress Dowager Chongqing in Birthday Celebration*

**Fig. 4.14** The Hall of Dispelling Clouds building complex and the Tower of Buddhist Incense

**Fig. 4.15** Restored birdview of Nimbus Land on Sumeru Mountain (Redrawn from the book *Summer Palace* by School of Architecture, Tsinghua University)

Qingyi Yuan. Its layout incorporated the characteristics of Han Buddhist temples, with multicolored glazed tiles and to manifest the imperial prestige. Emperor Qianlong originally intended to lay a raised platform at the rear of the temple to hold a Longevity Pagoda on top in the same way as the Six Harmonies Pagoda in Hangzhou, making it the composition center of the entire Longevity Hill.

But to his astonishment, in the 22nd year of Qianlong's reign (1757), when the emperor was happily expecting the completion of the Longevity Pagoda, the already-built eight-story pagoda was demolished for some unexpected reason that nobody knows. One speculation is that the huge pagoda overwhelmed the raised platform and continuation would lead to an accident. Another explanation is that as the work went on, the emperor discovered that the Dipamkara Pagoda at the top of the Jade Spring Hills was identical to the Longevity Pagoda in style, which may lead to negative visual effect. So he decided to change it into a three-story octagonal Buddhist tower we see today (Fig.4.14). The building was named Tower of Buddhist Incense, after the long Buddhist history of worshiping Buddha with incense. Back to back with the Temple of Immense Gratitude & Longevity

(not on the same axis) was another grand-sized Han-Tibetan Buddhist temple—Nimbus Land on Sumeru Mountain (Fig.4.15), which was a companion piece of the Temple of Universe Peace (Puning Temple) in Chengde. Compared with the Temple of Immense Gratitude & Longevity, although both temples were built on a hill, the greatest feature of this temple was that it was a built in imitation of the Samye Temple in Tibet.

4.13
4.14    4.15

A real-life Buddhist world was made up of the Belvedere of Fragrance Rock & True Essense at the center, surrounded by the Four Great Continents—the Eastern Continent (Purvavideha), Western Continent (Aparagodaniya), Northern Continent (Uttarakuru), Southern Continent (Jambudvipa), and Eight Subordinate Continents, together with the Sun Terrace, Moon Terrace, and the Four Colour Stupas symbolizing 4 wisdoms. The "Sumeru Mountain" stood at the center, surrounded by 4 different dimensions of world encircled by vast seas. Besides, two smaller temples, the Clouds Gathering Temple and Sudarsana Temple, guarded the main buildings from the west and east. Obviously, the Longevity Hill was just like a "Buddhist hill" or "sacred hill".

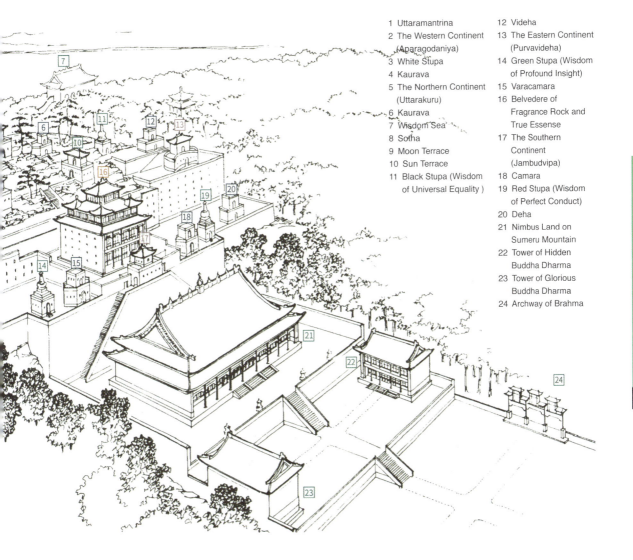

1 Uttaramantrina
2 The Western Continent (Aparagodaniya)
3 White Stupa
4 Kaurava
5 The Northern Continent (Uttarakuru)
6 Kaurava
7 Wisdom Sea
8 Sotha
9 Moon Terrace
10 Sun Terrace
11 Black Stupa (Wisdom of Universal Equality )
12 Videha
13 The Eastern Continent (Purvavideha)
14 Green Stupa (Wisdom of Profound Insight)
15 Varacamara
16 Belvedere of Fragrance Rock and True Essense
17 The Southern Continent (Jambudvipa)
18 Camara
19 Red Stupa (Wisdom of Perfect Conduct)
20 Deha
21 Nimbus Land on Sumeru Mountain
22 Tower of Hidden Buddha Dharma
23 Tower of Glorious Buddha Dharma
24 Archway of Brahma

## Pleasure scenes at the Longevity Hill

Compared with the religious buildings occupying the composition center, the other scenic spots disposed by Emperor Qianlong at the Longevity Hill appeared to be fairly "restrained", probably because he knew that building this garden had broken his previous promise.

As Qingyi Yuan was merely 1 km away from the Gate of Algae Garden of Yuanming Yuan, Emperor Qianlong did not plan for either himself or his mother to reside there. He simply set the east-facing palace gate and a small number of palatial buildings like the Hall of Diligence in State Affairs (now the Hall of Benevolence and Longevity), Pleasant Spring Hall (now Garden of Virtue and Harmony), Jade Billows Hall and Hall of Happiness & Longevity, according to the rank of imperial travel palace like Jingming Yuan.

The Gorgeous Sunset Tower is a belvedere on the west side in the second courtyard of the Jade Billows Hall (Fig.4.16). Here one can look west to the sunset and the beautiful lakes and hills of the Three Hills and Three Gardens. Its name comes from a line of Tao Yuanming. However, to punish himself for breaking his word, the emperor decided that he had to leave before midday and could not stay there overnight. So he had no chance of appreciating the sunset there.

After walking past the Hall of Happiness & Longevity, one would see the main mountain-front tour route. Stretching from the Gate of Inviting Moon in the east to the Pavilion of Saluting Rock❶ in the west. It measured 728 m in length. The route consisted of 273 bays of corridor, looking quite peculiar in the THFG (Fig.4.17). Although the Long Corridor was installed by the side of the lake, all scenic spots of the Front Hill were associated with it in layout. Apart from the east and west, there were four pavilions and two water pavilions in the middle of the corridor. Their NS-trending axis line was also the axis line of a succession of scenic spots including the Strolling in Paintings, Mansion of Listening to Orioles (Fig.4.18), Nest in Clouds & Pines, and Pavilion of Depicting Autumn. From this one can see how precise

❶ The names comes from the story of Mi Fu saluting the rock.

Fig. 4.16
The Gorgeous Sunset Tower on the eastern bank of the Kunming Lake

Fig. 4.17
The Long Corridor at Yihe Yuan

Fig. 4.18
The Mansion of Listening to Orioles and Strolling in Paintings

Fig. 4.19
The rocky arrangement at the entrance to the Nest in Clouds and Pine

the overall layout of the Longevity Hill was. These small gardens were not large in volume and were quite regularly arranged. But as they were built according to the hill, they created a rich vertical variation and profound cultural connotation. The Nest in Clouds & Pine and Shao Yong's Nest were two typical examples (Fig.4.19). "Nest in Clouds and Pine" took its name from a line of the famous Tang poet Li Bai and this phrase meant the place of seclusion. "Shao Yong's Nest" directly declared that it was named in honor of the well-known North Song Neo-Confucianist Shao Yong (1011—1077). Using "nest" to name a palace also carries a very obvious sign of seclusion from the earthly world. The names of the scenic spots, such as "Pavilion of Depicting Autumn" "Strolling in Paintings" "Natural Interest of Lakes and

**Fig. 4.20**
Ruins of the Watching Clouds Rising and Rear Stream River

**Fig. 4.21**
The rebuilt Suzhou Street

**Fig. 4.22**
Old photograph of the Garden of All-inclusive Spring cum Studio of Enjoying Leisure before they were ruined (1877)

Mountains" and "Relaxed Mood as Floating Clouds", expressed the emperor's appreciation of natural climate, seasonal changes and lake-hill views. Among these scenes were irregularly shaped buildings like the fan-like Promoting the Ethos of Benevolence and boat-like Marble Boat, enhancing the aesthetic taste of the garden.

Unlike the Front Hill, the scenic spots on the Rear Hill were quite disperse. Their orientation was highly variable because of the terrain. Their layout was quite flexible, too, giving full play to the features of gardens in the mountains. The Gorgeous View Pavilion and the Watching the Clouds Rising (Fig.4.20), Suzhou Street ❷ (Fig.4.21), Simplicity & Tranquility Hall and Huishan Garden ❸ were disposed along the deep, zigzag Rear-Stream River. Some of them faced each other across the water; some had their courtyards fixed on the river banks; some were secluded in the valleys away from the outside world. Garden of All-inclusive Spring (Fig.4.22) cum the Studio of Enjoying Leisure, Towering Pavilion, Blossom Clustering Belvedere and Six Pleasures Studio ❹ were seated halfway of the hill. Some of them consisted of layers upon layers of terraces for watching the creeks inside and farmland outside; some comprised raised platforms for meditating on Buddhist dharma. Despite their divergent physical locations and themes, these scenic spots had one thing in common: they each selected the representative landform of the Longevity Hill (e.g., exposed rocks) and remodeled or built houses there, so that the viewer would be able to cultivate their temperament while climbing up the hill. This reflects the sophisticated garden-building concept and construction workmanship of the ancient people.

❷ A water market built in imitation of downtown streets in Suzhou.

❸ Built in imitation of Jichang Yuan in Wuxi, renamed "Garden of Harmonious Interests" in Jiaqing period.

❹ The name signifies having all six good matters: good timing, beautiful scenery, pleasant mood, delightful event, hospitable host, distinguished quest.

1 Garden of All-inclusive Spring
2 Natural Interest and Pleasure
3 Gallery of Tranquility and Pleasure
4 Bamboo Chamber
5 Bell Pavilion
6 Chamber of Fragrant Rock
7 Lingering Cloud
8 Pavillion of Enjoying Leisure
9 Peach Blossom Ravine

Restored Plan

## Pleasure scenes on the Kunming Lake

The 3 islets in the lake continue with the ancient "One-pond Three-mountains" garden-building tradition. Unlike those in Yuanming Yuan or Jingming Yuan that are centralized together, these islets each occupies a separate part of the Kunming Lake, functioning as the composition center of their respective zone.

Before the garden was built, the South Lake Island on the east side used to be a Dragon King Temple (Temple of Extensive Moisture and Timely Rains) built to pray for favorite weather and peace of water. In order to both retain this temple and broaden the lake surface, the location of the temple was reworked into an island and linked to the East Causeway with an over 150 m long stone arch bridge—the Seventeen-arch Bridge. Emperor Qianlong commended this bridge as a "**slender rainbow across the waves**". At the end of the bridge, there was a huge double-eaved octagonal pavilion, the Spacious Pavilion (possibly a marker for the secondary entrance to the scenic area). A Bronze Ox for stabilizing waters was placed by the water (Fig.4.23). Such a tradition was used to maintain the perpetual peace and bring blessings to the Kunming Lake. Emperor Qianlong's article was also inscribed on the bronze ox.

Fig. 4.23
The Bronze Ox and the Seventeen-arch Bridge on the Eastern Causeway of the Kunming Lake

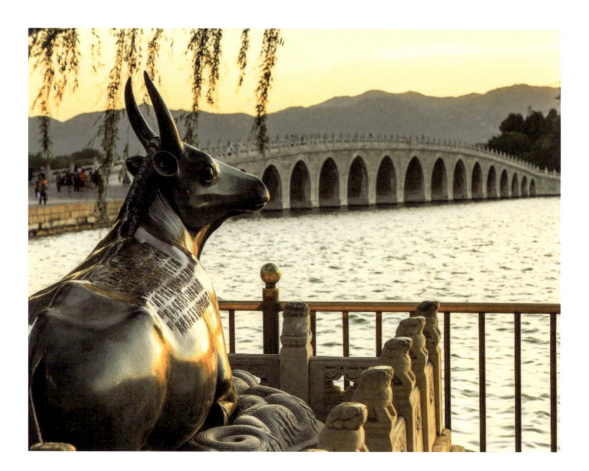

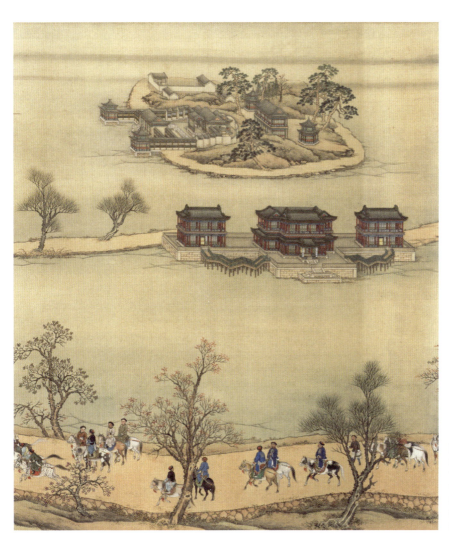

Fig. 4.24
Qing. Court Painter.
The Hall of Appraising
Talents in the *A Panorama
of Empress Dowager
Chongqing
in Birthday Celebration*

West of the West Causeway, the Belvedere of Mirroring State Affairs, known as the "Round City", sat on the towering double-layer circular city wall. On the periphery, there was a circular corridor and hall facing four directions. This plan layout displayed the mandaia ❺ style of the Tibetan Buddhism. It represented a second "real-life" Buddhist in addition to the Nimbus Land on Sumeru Mountain at the Longevity Hill. On the elliptical island to the south stood the Hall of Appraising Talents (Fig.4.24). This was a courtyard embraced by earth mounds, with a pavilion reaching down into the water on both its flanks. The pavilion at the back of the courtyard was called the "Terrace of Tasting Tea in Spring Breeze", which was a great location for savoring tea in Qingyi Yuan following the Hill Chamber with Bamboo Tea Stove at Jingming Yuan and the Study with Bamboo Tea Stove at Jingyi Yuan. Obviously, the 3 islands were shaped differently to reflect different artistic images.

❺ Chinese translation of Sanscrit Mandaia, meaning the palace inhabited by Buddha or Bodhisattva.

# 4.3 Ready in shape

In the 26th year of Qianlong's reign (1761), Qingyi Yuan was formally completed. This not only signified the completion of an imperial garden, but also a successful modification and upgrade to the visage and functionality of the entire northwestern suburbs, drawing a full stop to the overall composition of the THFG. Over this prolonged period of 12 years, it was under the general control of Emperor Qianlong that a perfect man-nature integration was accomplished in the western suburbs. Here we can feel his rational spirit of being concerned about farming and water conservancy, his religious stance of remaining filial to his mother and pious to Buddha, and his literary aesthetics of finding emotional echoes from hills and waters. Not only did he push the Chinese ancient landscape architecture up to a new high, he also made great contribution to the ecological protection in the northwestern suburbs of Beijing.

The construction of Qingyi Yuan and the water conservancy infrastructure helped connect and integrate the "Three Hills" both spatially and visually. The aqueducts at the Western Hills connected the Fragrant Hills to the Jade Spring Hills. The serpentine Jade River connected the Jade Spring Hills with the Longevity Hill. The Kunming Lake supplied water to the paddy fields in Haidian and gardens like Yuanming Yuan. This way, the plain palatial garden district represented by Changchun Yuan and Yuanming Yuan in the east were truly combined with the upland palatial district represented by the Three Hills and Three Gardens in the west. This signifies that by virtue of a total oneness, the THFG region constituted a special administrative area of the Qing empire on the outskirts of the capital.

4.25

4.26

**Fig. 4.25**
Seen westward from the Longevity Hill cum Yihe Yuan

**Fig. 4.26**
Seen eastward from the Fragrant Hills cum Jingyi Yuan

    In terms of visual appreciation, the diversely-styled pagodas or belvederes towering at the Three Hills made up a number of commanding heights and architectural icons in the western suburbs. Against the backdrop of the magnificent Western Hills, they compounded a perfect "natural picture" together with the Tibetan blockhouses and temple buildings on the periphery of the Fragrant Hills, as well as the colored plant landscapes imitating those beyond the Great Wall. Today, whether we look west from the eastern bank of the Kunming Lake, or east from the Fragrant Hills, we will obviously feel an air of solemnity combining the imperial majesty with Buddhist wonderland, as well as an overall visual impact of gradation and rhythm (Fig.4.25-Fig.4.26).

# 4.4 From Qingyi to Yihe

**In this section, you will get an idea about:**
1. What are the differences between Qingyi Yuan and the rebuilt Yihe Yuan?
2. Why is it incorrect to call this garden Summer Palace?

With the decaying national power of the Qing empire, the imperial gardens did not look as nice as they had been before. In the late years of Daoguang period, the emperor ceased touring the Three Hills (the tour was resumed in the Xianfeng period) and withdrew a lot of the imperial furnishings from the garden.

In the 10th year of Xianfeng's reign (1860), The vast majority of the scenic spots in Qingyi Yuan were burned down.

In the 13th year of Tongzhi's reign (1874), the attempt to renovate Yuanming Yuan ended in a failure. Months later, work on the Three Seas of the West Palace was also forced to stop. Yet Empress Dowager Cixi had not totally given up the "obsession" of renovating the garden. Her arrogation of all powers to herself ought to have ended at the time when Emperor Guangxu turned 18 years old (1889), but it continued until she died in 1908. Yihe Yuan was born in the 12th year of Guangxu's reign (1886), right before the year in which she was to return the ruling power back to Emperor Guangxu as she had originally planned.

That year, the Kunming Lake Navy Academy was built up in Qingyi Yuan at the location of the former Weaving and Dyeing Bureau. Two years later, people were surprised to realize that the ruins of Qingyi Yuan having lain waste for nearly 30 years was reused for some other purpose. From the mouth of Emperor Guangxu, Empress Dowager Cixi found a new excuse for renovating the garden: The emperor wanted to promote filial piety and wish her mother a long life, like what Emperor Qianlong had done for his own mother. He also declared that Qingyi Yuan was renamed "Yihe Yuan (Garden of Recuperating Harmony)", just to emphasize good blessings to his mother for happiness and longevity. The construction of Yihe Yuan cost more than 5 million liang Silver. The money did not come out of the imperial treasury, but was directly taken from the naval construction funds ❻. In 1894, just when people were busy preparing for the ceremony celebrating Empress Dowager Cixi's 60th birthday in Yihe Yuan, the Sino-Japanese War broke out. The war ended up with the disastrous defeat of the Qing navy. In the same year, the Navy Administration was dissolved and work on Yihe Yuan was suspended.

❻ As originally planned, 2.6 million liang Silver was to be pooled among high-rank officials around China and deposited in the navy's funds. The interest accrued would be used to renovate the garden. Unfortunately, the money was not fully pooled in time and the interest was far from sufficient. They had to use military funds and tariffs.

After almost 10 years of work, Yihe Yuan mostly revitalized the original appearance of Qingyi Yuan. The vast majority of the scenic spots at the Front Hill and Kunming Lake had been either renovated or reconstructed. The difference was that the plaques and couplets hung on the buildings were largely different from those in Qianlong period. They are filled with gorgeous wordings in praise of merits and peace.

The Temple of Immense Gratitude & Longevity in the south of the Longevity Hill was modified into a Hall of Dispelling Clouds building complex and no longer served as a temple. Its layout looked more like a "transplant" of the outer court in the Forbidden City. The Tower of Buddhist Incense was rebuilt as it used to be. To cater for daily life and amusement, the proportion of living and performance spaces was increased near the Hall of Happiness & Longevity. The Jade Billows Hall and Mansion of Book Collection were used as the bedchambers of the emperor and his empress and concubines. The empress dowager herself resided in the lake-front Hall of Happiness & Longevity (Fig.4.27-Fig.4.28), which was accessible by the Wharf of Natural Affinity of Water & Woods. The Grand Opera Tower at Garden of Virtue & Harmony became a performance center for palace dramas in the late Qing dynasty (Fig.4.29-Fig.4.30). The Imperial Opera Department was moved from outside Yuanming Yuan to outside Yihe Yuan. Garden of Harmonious Interests (Fig.4.31) and the Pavilion of Freshness after Snow at the east piedmont of the Longevity Hill were also renovated. They even served as a location for temporary administration of state affairs.

**Fig. 4.27**
The Wharf of Natural Affinity of Water & Woods

**Fig. 4.28**
The Hall of Happiness and Longevity

　　Besides, many scenic spots projected to be rebuilt were either canceled or their plans were simplified due to short budget. For example, the Wenchang Emperor Belvedere on the eastern bank of the Kunming Lake, Epiphyllum Belvedere (i.e., the Utmost Blessing Belvedere at Yihe Yuan) at the Longevity Hill, and the Belvedere of Fragrance Rock & True Essense at the Nimbus Land on Sumeru Mountain all had their height reduced. The Marble Boat beside the Kunming Lake was turned from a Chinese-style tower into the Sino-European "hybrid" and renamed "Clear and Peaceful Boat" (Fig.4.32). Pitifully, the Suzhou Street, Towering Pavilion, Garden of All-inclusive Spring, and the Simplicity &

Tranquility Hall on the Rear Hill of the Longevity Hill were discarded. More than that, for the sake of the safety of the imperial family, the coverage of Yihe Yuan was also far greater than Qingyi Yuan (which was originally limited to the Longevity Hill area). The entire Kunming Lake was enclosed into the bounding wall. The bounding wall itself was also heightened.

Shamefully, this garden was not spared by the looting and destruction of the Eight-Power Allied Forces. In the 7th month of the 26th year of Guangxu's reign (1900), the Allied Forces arrived at the capital city. Empress Dowager Cixi, together with Emperor Guangxu, hurried to Yihe Yuan from the Forbidden City and headed for Xi'an soon afterwards. The foreign armies flocked in and used Yihe Yuan as their campsite. They did not withdraw until the Qing government signed a *Boxer Protocol* with the powers the next year. During the last few years of the Qing dynasty, Yihe Yuan, as a diplomatic venue, was highly famous in the world. It was called the (New) Summer Palace in comparison with Yuanming Yuan (which is called the Old Summer Palace). But this place was not at all intended as a summer resort.

On the 26th day of the 9th month of the 34th year of Guangxu's reign (October 20, 1908), Empress Dowager Cixi and Emperor Guangxu left Yihe Yuan, never to come back again. A month later, both died at the Middle and South Sea. Yihe Yuan was sealed. In 1912, the issuance of an imperial edict of abdication by Emperor Xuantong declared the doom of the Qing dynasty. Yihe Yuan became the last imperial garden in the Chinese history (Fig.4.33).

**Fig. 4.29**
Qing. Style House. Hot model of Garden of Virtue and Harmony

**Fig. 4.30**
Wu Xiaoping. The Grand Opera Tower at the Garden of Virtue and Harmony

**Fig. 4.31**
Garden of Harmonious Interests at Yihe Yuan

**Fig. 4.32**
Marble Boat (Clear and Peaceful Boat) at the northwestern of the Kunming Lake

1. Graceful Scenery Archway
2. East Palace Gate
3. Hall of Benevolence and Longevity
4. Wenchang Gallery
5. East Gate of Wenchang Gallery
6. Yelyu Chucai Shrine
7. Wenchang Emperor Belvedere
8. Pavilion of Heralding Spring
9. Jade Billows Hall
10. Gorgeous Sunset Tower
11. Mansion of Book Collection
12. Garden of Virtue and Harmony
13. East Eight Offices
14. Tea Serving House
15. Gateway of Purple Cloud from the East
16. Natural Affinity of Water and Woods
17. Hall of Happiness and Longevity
18. Promoting the Ethos of Benevolence
19. Pavilion of Containing Freshness
20. Pavilion in Woods and Clouds
21. Pavilion of Cultivating Clouds
22. Gateway of Colorful Peaks
23. Dense Green Pavilion
24. Blessed Fortune Pavilion
25. Relaxed Mood as Floating Clouds
26. Gazebo of Facing Waterfowl
27. Endless Charm Pavilion
28. Pavilion of Depicting Autumn
29. Extended Longevity Hall
30. Hall of Dispelling Clouds Building Complex
31. Pavilion of Clear Water and Flourishing Woods
32. Revolving Archives
33. Tower of Buddhist Incense
34. Ratnamegha Bronze Pavilion
35. Shao Yong's Nest
36. Nest for Pines in Clouds
37. Pavilion of Fish and Algae
38. Tower of Panoramic View of Lakes and Mountains
39. Endless Wealth and Longevity
40. Mansion of Listening to Orioles
41. Strolling in Paintings
42. Natural Interest of Lakes and Mountains
43. Pavilion of Saluting Rock
44. Pavilion of Recepting Blessings
45. West Four Offices
46. Hall of Placing Emotions on Billows
47. Clear and Peaceful Boat(Marble Boat)
48. Waterpoppy Bridge
49. Five Saints Shrine
50. Tower of Greeting the Sunrise
51. Imitated Xiling
52. Clearing Heart Belvedere
53. Riverside Hall
54. Tower of Lingering Elegant Vision
55. Taoist Fairyland
56. Gateway of Cloud-retaining Eaves
57. North Dock
58. Half-wall Bridge
59. Hall of Rejuvenating Virtue
60. Ruyi Gate
61. Gateway to the Misty Fairyland
62. Temple of Buddha's Perfect Practice
63. Painted Blossoms Hall
64. Suzhou Street
65. North Palace Gate
66. Big Screen Wall
67. Clouds Gathering Temple
68. Nimbus Land on Sumeru Mountain
69. Sudarsana Temple
70. Gateway of Dawn Light
71. Glazed Pagoda
72. Simplicity and Tranquility Hall
73. Utmost Blessing Belvedere
74. Prolonged Longevity Hall
75. Satisfaction Villa
76. Pavilion of Happiness in Farming
77. Garden of Harmonious Interests
78. Pavilion of Freshness after Snow
79. Northeast Gate
80. Studio of Overlooking Faraway
81. Imperial Opera Department
82. Lake Boundary Bridge
83. Waterside Dwelling
84. Naval Academy
85. Studio of Lingering Appreciation
86. Imperial Stone Tablet of Pictures of Plowing and Weaving
87. Binfeng Bridge
88. Jade Belt Bridge

# Current plan of the Longevity Hill at Yihe Yuan (2020) Fig.4.33

# Touring the Three Hills and Five Gardens

Landscape, Art and Life of
Chinese Imperial Gardens

# Chapter 5
# Garden-based Life of the Imperial Family

For a long time, people have tended to assume that the Qing emperors and their families were kept almost all year round within the tall bounding walls of the Forbidden City; they had nowhere to relax themselves except the Imperial Garden inside there. In reality, most of their time was spent in the THFG, since it possessed all necessary functions plus beautiful sceneries. People hold this misperception mainly because most of the gardens in the THFG region are no longer existent as physical beings; efforts employed in history mining and dissemination also need to be further enhanced.

When they did exist, the gardens each had their own particular functions. They possessed dizzying varieties of themes and scenic spots, and contained distinctive scenes and experiences across different seasons. These greatly enriched the material and spiritual lives of the imperial family, making the quality of their life incomparable for anyone else. In Qianlong period, Changchun Yuan, the garden where the empress dowager resided, covered approximately 0.78 times the size of the Forbidden City. Yuanming Yuan (without accounting for the accessory gardens), the garden resided by the emperor and his empress, concubines and princes, was 2.8 times as large as the Forbidden City. Charngchun Yuan, the retirement garden of Emperor Qianlong himself, was basically as large as the Forbidden City. Obviously, such enormous imperial gardens were not covered with palaces and plazas like the Forbidden City. If we compare a garden to the human body, the hills and waters in it should be the bones and blood veins covering the larger part of the garden. The plants should be the skins to guarantee a comfortable ecological environment. The buildings should be the organs and muscles to provide concrete dwelling and pleasure functions. In other words, as the imperial palaces and imperial gardens were planned under essentially divergent lines of thought, their dwelling experiences were poles apart.

If we want to know how much the Qing monarchs enjoyed residing in gardens, this time schedule of the Qing emperors' residing in Yuanming Yuan will provide the best evidence. According to authoritative experts in this field, since Emperor Yongzheng formally settled down in Yuanming Yuan in the 3rd year of Yongzheng's reign (1725), over the next 10 years, he had stayed in the garden for an average of 206.8 days a year. Since Emperor Qianlong formally settled down in Yuanming Yuan in the 3rd year of Qianlong's reign (1738), over the next 61 years, he had stayed in the garden for an average of 126.6 days per year. One year, Emperor Qianlong stayed in the garden for 251 days. The falling number of days was attributable to his frequent inspection tours away from Beijing. Since Emperor Jiaqing formally settled down in Yuanming Yuan in the 1st year of Jiaqing's reign (1796), over the next 25 years, he had stayed in the garden for an average of 162 days a year. The number of days spent by Emperor Jiaqing in the garden ranged from 111 days to 247 days. In Daoguang and Xianfeng periods, the emperors recorded the longest numbers of days spent in Yuanming Yuan. They spent an average of 260.1 days a year and 216.4 days a year, respectively. In the 29th year of Daoguang's reign, the emperor stayed in Yuanming Yuan for 355 days. That was the longest time across the five periods.

That being said, what kind of a role did these countless garden scenes play in the palatial life in real history? How was the life of the emperors and their empress and concubines in these gardens? Definitely you will find the answers from this chapter.

# 5.1 Court at front, life at rear

**In this section, you will get an idea about:**
1. How did the state organs of the Qing dynasty operate in the THFG?
2. What characterized the work and dwelling places in Yuanming Yuan?
3. In what locations did the emperor's empress, concubines and princes reside in Yuanming Yuan?

In the 26th year of Kangxi's reign (1687), in Changchun Yuan, the emperor inaugurated constant "garden-based governance" in the imperial gardens in the western suburbs. The ministers had no other choice but to follow him into the THFG away from the capital city. How was everything going on then?

## Establishment of the rotational statement regime

Before Changchun Yuan was completed, Emperor Kangxi would come for a short stay at the Jade Spring Hills cum Jingming Yuan once in a while. On those occasions he would meet with his ministers at the Front Pavilion every day. Obviously, even a simple pavilion in a garden could serve the same function as the Gate of Heavenly Purity in the Forbidden City. Here in Changchun Yuan, the places for attending state affairs included the Simplicity & Tranquility Hall and the quadrangle dwelling in front of it. Nearby, a duty place was also set for the imperial academicians of the South Study to answer to the emperor whenever he called. This routine went on until the 52nd year of Kangxi's reign (1713), when the Hall of Nine Classics & Three Events appeared in Changchun Yuan for holding major ceremonies. By then, it was almost three decades since Changchun Yuan was initially built. The emperor was obviously very restrained over garden building.

After Emperor Yongzheng succeeded to the throne, he devoted all his mind to creating his ideal imperial garden—Yuanming Yuan. Drawing on experiences from Changchun Yuan, the emperor prepared full-functional office work spaces for himself and his ministers. He also planned comfortable living spaces for his family. In the 3rd year of Yongzheng's reign (1725), after he had just formally resettled himself in Yuanming Yuan, Emperor Yongzheng issued an edict, "**I feel nothing different here in Yuanming Yuan than back in the imperial palace. Anything that must be done shall be done as usual. If you have something to report to me, just do it without delay. If you don't have anything to report, you don't need to come here. Just do what you must do at your own office.**"

He also established a well-defined rotational statement regime: the Eight Banners would each take turns for 1 day; among the government departments, the Ministry of Personnel, Ministry of Revenue and Population, Ministry of Rites, Ministry of Military, Ministry of Penalty, and Ministry of Engineering would take turns for 1 day each; the Court of Supervision and Court of Vassal State Affair would take 1 day; the Imperial Household Department would take 1 day. This way, 8 days would be a cycle. If there was something urgent, a statement should be made right away rather than waiting until the right day.

Emperor Yongzheng worked from dawn to dusk, racking his brains for the sake of the Qing Empire. He was still working even 2 days before he passed away. Regarding the cause of his sudden death in Yuanming Yuan, some suggested that the emperor was trying to extend his lifespan by taking an elixir containing heavy metals, which eventually took his life.

Each year, before the 15th of the first month, Emperor Qianlong would normally come back to reside in Yuanming Yuan with his own family. This way, besides daily statement in the court, major ceremonies such as diplomatic receptions and imperial competitive examinations were also held in Yuanming Yuan. Gradually, the THFG where Yuanming Yuan is located became a political center that is equally important to, and even a little more important than, the Forbidden City.

Such a regime was not free of imperfection. In Jiaqing period, the state administrative departments back in the capital city were found to very inactive. Hence, some ministers tried to persuade the emperor to suspend residing in the imperial garden. But these suggestions were harshly rejected by the emperor.

Next, let us know each of the office and dwelling spaces of the emperors.

### The Grand Palace Gate of Yuanming Yuan

The main entrance of Yuanming Yuan was magnificently structured. The outer plaza alone was 43,000 m² in area and about 300 m in length. After passing the Fan Lakes, the imperial road directly led to the center of the plaza, forming a T-shaped road fork at this point. Further north along the central axis sat the Grand Palace Gate guarded by two brass lions. On the southern part of the plaza lay a serpentine guarding river and a 40 m long Big Screen Wall. The entire plaza was surrounded by wooden fence, pinpointing the imperial forbidden area (Fig.5.1).

Between the Grand Palace Gate and Secondary Palace Gate, namely, the Gate of Virtuous (1# in Fig.5.2), was the inner plaza, which was less than 7,000 m² in area and only 75 m in length. Despite its serious appearance, the place was no lack of life. The bow-like golden stream and few trees dotted in the courtyard both gave an impression of affinity. Besides, all buildings here were composed of lower-grade round-ridge roofing and gray tiling, rather than the golden-colored glazed tiles found everywhere in the Forbidden City. The L-shaped corner Reception Rowhouse for Officials and Reception House for Officials at the two palace gates were places where the ministers worked their shift and waited for meeting with the emperor. Some places may have felt a little crowded when two or more departments shared the same room. But after all, these were not intended for daily office work. This area also housed service buildings like kitchens and tea houses.

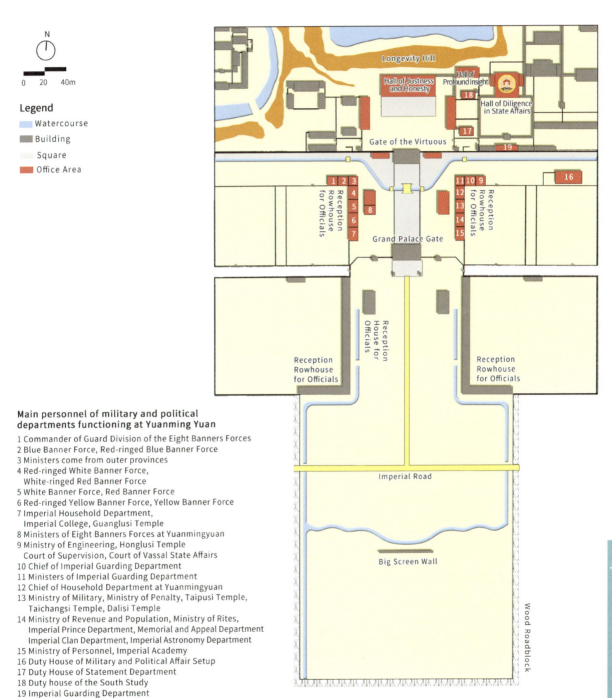

**Main personnel of military and political departments functioning at Yuanming Yuan**

1 Commander of Guard Division of the Eight Banners Forces
2 Blue Banner Force, Red-ringed Blue Banner Force
3 Ministers come from outer provinces
4 Red-ringed White Banner Force,
  White-ringed Red Banner Force
5 White Banner Force, Red Banner Force
6 Red-ringed Yellow Banner Force, Yellow Banner Force
7 Imperial Household Department,
  Imperial College, Guanglusi Temple
8 Ministers of Eight Banners Forces at Yuanmingyuan
9 Ministry of Engineering, Honglusi Temple
  Court of Supervision, Court of Vassal State Affairs
10 Chief of Imperial Guarding Department
11 Ministers of Imperial Guarding Department
12 Chief of Household Department at Yuanmingyuan
13 Ministry of Military, Ministry of Penalty, Taipusi Temple,
   Taichangsi Temple, Dalisi Temple
14 Ministry of Revenue and Population, Ministry of Rites,
   Imperial Prince Department, Memorial and Appeal Department
   Imperial Clan Department, Imperial Astronomy Department
15 Ministry of Personnel, Imperial Academy
16 Duty House of Military and Political Affair Setup
17 Duty House of Statement Department
18 Duty house of the South Study
19 Imperial Guarding Department

# Restored plan of the Grand Palace Gate area of Yuanming Yuan  Fig.5.1

## The Hall of Justness and Honesty

At the terminal end of the courtyard compound stood the main hall of Yuanming Yuan. Seated on an elevated terrace, it consisted of 7 bays with surrounding corridors. The main hall (2# in Fig.5.2) measured 36.45 m across by 16 m deep, with a capacious platform at the front. Functionally, this hall was equivalent to the Hall of Supreme Harmony or the Hall of Preserving Harmony. It was intended for attending state affairs, feting vassal states, interviewing with diplomatic envoys, and holding imperial examinations. The other two features of the Hall of Justness & Honesty were the huge birdview of Yuanming Yuan on the western wall inside the room and the high-rising "Swordlike Rocks" behind the hall that symbolized the supremacy of imperial power (Fig.5.2).

## The Hall of Diligence in State Affairs

East of the Hall of Justness & Honesty lay the office space cored around the Hall of Diligence in State Affairs and Hall of Profound Insight (3# in Fig.5.2). The three buildings were connected to one another by corridors to make it convenient for the emperor to move between the buildings. The Statement Department and the duty room of South Study[1] stood immediately west of the Hall of Diligence in State Affairs. The Imperial Guarding Department was seated south of the Hall of Diligence in State Affairs. People working there were all favorite ministers to the emperor. The name Hall of Profound Insight means clear insights. As in the Qing dynasty, all prisoners on death row were to be approved by the emperor himself, hence here was a location for determining whether criminals were to live or die.

The Hall of Diligence in State Affairs was quite similar to the Simplicity & Tranquility Hall at Changchun Yuan in layout and functions. Both were quadrangle dwellings. The Hall of Diligence in State Affairs was not big itself, with only three bays. In Xianfeng period, it measured 11.8 m across by 7 m deep (Fig.5.3). The inside of the hall was divided into a front and a rear part. The front part was larger and intended for the ministers; the rear part of smaller and intended for the emperor himself. Opening the northern window, one would see the gracefully-shaped rockeries of Taihu Rock. Although the Hall of Diligence in State Affairs was functionally comparable to the Hall of Mental Cultivation at the Forbidden City, its layout and decoration were very simple, just to carry forth the family virtue of simplicity and frugality. Hence it was solely a small office room.

## The Depths of Fairy Caves

Starting from Changchun Yuan, raising princes was also an important function of imperial gardens. The West Garden, and the Studio of Working without Ease at Changchun Yuan, used to serve the dwelling and schooling of the princes in earlier years, respectively. In Yuanming Yuan, these functions were combined into one single scenic area. The Depths of Fairy Caves was encircled by earth mounds and bounding walls (Fig.5.4). It could be accessed from the Gate of Blessing & Affinity or by a

[1] The place where the literature attendants of the Qing emperor served their duty.

**Fig. 5.2** Qing. Court Painter. The Hall of Justness and Honesty in the *Album of Paintings and Poems of Forty-Scenes of Yuanming Yuan*

**Fig. 5.3** Restored interior plan of the Hall of Diligence in State Affairs (Xianfeng period) (Redrawn from the book *Exploration of Gardening Art of Yuanming Yuan* by Jia Jun)

**Fig. 5.4** Qing. Court Painter. Part of the Depths of Fairy Caves in the *Album of Paintings and Poems of Forty-Scenes of Yuanming Yuan*

boat cruise from the water gate at the northern bounding wall. Upon entry into this area, one would feel like entering a "fairy cave" of the Taoist immortals, showing the monarch's eagerness to pursue immortality.

The names of the halls show the expectations of Emperor Yongzheng. "Congenital Grace" (1# in Fig.5.4), "Culminant Prosperity" (2# in Fig.5.4), and "Future Immortality" (3# in Fig.5.4) mean gods blessings in the pre-heaven, flourishing prosperity in the mid-heaven, and living forever in the post-heaven. He selected this place as the schooling site for his sons, possibly because he wanted to infuse this idea into the youngsters: They were highbred and would live up to this privilege only by working hard and creating a flourishing age. Here one could also find a Saint's Hall for worshiping Confucius and a waterfront Fairy Terrace (4# in Fig.5.4). Standing up there, one would feel like having a conversation with heaven.

The Four Dwelling Palaces (5-8# in Fig.5.4) for Princes lay east of the scenic area. It consisted of 4 totally identical, large three-courtyard quadrangle dwellings. The emperor was obviously even-handed for all his sons. Later on, Emperor Daoguang combined the

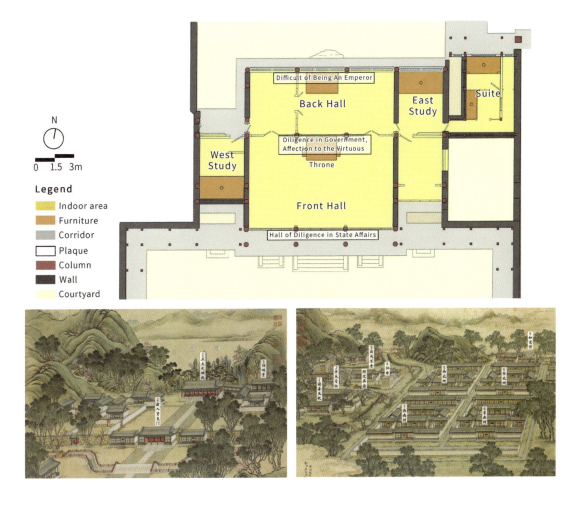

northern and southern courtyards into two dwellings running through the north and south. The courtyards were made even deeper and quieter.

The Well-content Mansion (9# in Fig.5.4) sat in the northeast of the Four Dwelling Palaces for Princes. The pavilion was rustically composed of 3 small houses, yet many well-known Chinese and foreign painters in history have worked there. Emperor Qianlong, with his high level of artistic taste, often came to inspect the place.

## The Peace over the Nine Prefectures

The Peace over the Nine Prefectures was oldest scenic spot and bedchamber for the emperor and his empress and concubines (Fig.5.5). On this huge, 22,000 m² island, there were 20 palaces and 30 accessory houses. In fact, it was the scenic area boasting largest scale and the greatest modification frequency among all other scenic areas in Yuanming Yuan. The island was linked to the outer court area by two bridges. Access was permitted for nobody else other than the imperial family members and servants.

The building complex was divided into a central part, an eastern part, and a western part. The high-status "three halls" in the middle served as a gate hall, a banquet hall, and a bedchamber. This layout had been maintained all the time. The eastern part was cored around the large quadrangle dwelling named "Gathering Spring in Heaven and Earth". The quadrangle dwelling was home to the bedchambers of the many concubines as well as the dwelling or duty rooms for the maidservants and eunuchs. The western part was the residing and living space of the emperor. It covered almost a half of the area of the island. In Daoguang and Xianfeng periods, great changes had already taken place. The main differences included the huge emperor's bedchamber named the "Hall of Cultivating Virtue", and the spacious front yard.

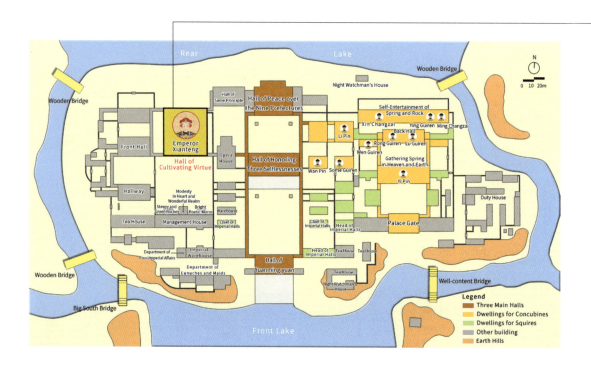

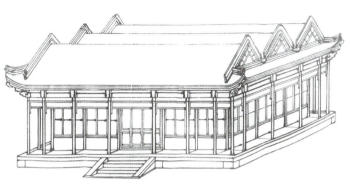

**Fig. 5.5**
Restored plan of imperial resting palaces in the Peace over the Nine Prefectures (Xianfeng Period)

**Fig. 5.6**
Qing, Style House. *Axonometric Drawing of the Hall of Cultivating Virtue*

**Fig. 5.7**
Restored interior plan of the Hall of Cultivating Virtue (Xianfeng Period) (Redrawn from the book *Exploration of Gardening Art of Yuanming Yuan* by Jia Jun)

These were reconstructions from the 11th year of Daoguang's reign (1831).

The Hall of Cultivating Virtue was one of the largest single buildings in Yuanming Yuan (Fig.5.6-Fig.5.7). Although it had only 5 bays, its depth was actually comprised of three buildings spliced together. The hall measured 19.1 m across by 22.24 m deep. Its area was 424.8 m². However, the bulky building did not look hollow inside. It was flexibly partitioned into a number of spaces of varying sizes in the front, rear, left and right sections, looking like a labyrinth. Inside, there a stairway led to the two-story fairy tower.

The yard of the hall was no less marvelous. Covering about 800 m² of land, the yard was studded with a rich variety of ornamental plants such as peony, paeonia, apple, persimmon, Chinese pine, yacca, flowering plum and apricot and ornaments such as stone desks and stools. On the rockery opposite to the Hall of Cultivating Virtue stood three high-rising pavilions as the opposite scenes to the palace. Climbing up, one would get an overview of the entire Nine-Prefectures scenic area.

Like in the Forbidden City, following the established rules of their forefathers, the emperors did not reside in the same building as their empress and concubines. Lower-rank concubines mainly resided in the eastern portion of the Peace over the Nine Prefectures. During Xianfeng period, the courtyards in the Gathering Spring in Heaven & Earth compound were inhabited by 10 concubines, including Yi Pin (who later became Empress Dowager Cixi)(Fig.5.8), Li Pin, Rong Guiren, Ying Guiren, Xin Changzai, and Ming Changzai. Each of them was provided with accessory houses for their maidservants.

From the birthplace of Emperor Xianfeng, we can know that the empress of Daoguang once resided in the Studio of Imperturbation (later renamed Hall of the Foundation of Blessing) west of the Hall of Cultivating Virtue. From pictorial archives, the empress and higher-rank concubines must have lived inside the courtyard of the Hermit of Eternal Spring's Fairyland (Fig.5.9). This island covered an area of approximately 8,000 m². It used to be resided by Emperor Qianlong before his accession to the throne. Later on, it became the bedchamber of the Empress Dowager Chongqing for short stays

**Fig. 5.8**
Palace of Gathered Elegance of the Forbidden City where Cixi once lived

**Fig. 5.9**
Hermit Eternal Spring's Fairyland in the Album *of Paintinys and Poems of Forty-Scenes of Yuanming Yuan*

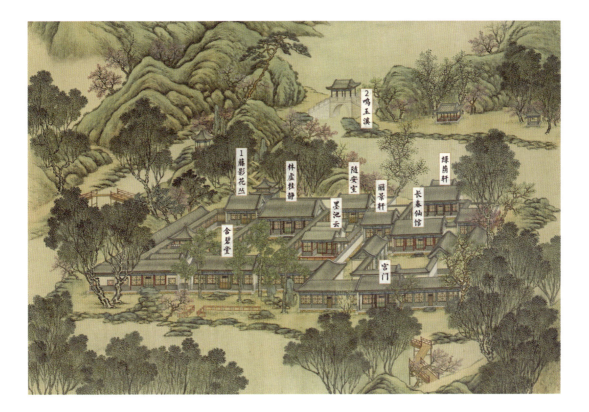

in Yuanming Yuan. It was also a temporary adobe for Emperor Jiaqing after his father abdicated the throne. Hence the place was of high importance. The empress resided in the Hall of Wistaria Shade & Blossoms (1# in Fig.5.9 ) at the westernmost side. This hall had a floor area of 170 m². Opening the back window of the hall, one would see a quadrilateral pavilion standing on the rockery outside the yard. On the earth hummock behind the pavilion, there was another pavilion. From there, one could look into the Stream of Knocking Jade across the water (2# in Fig.5.9 ). The maidservants of the empress resided in the 12 rooms west of the yard.

In order to provide 24-hour attentive catering and medical services for the imperial families, the Imperial Kitchen, Imperial Tea Department, and Imperial Medicine Department were located on the land immediately west of the Hermit Eternal Spring's Fairyland. Here, the catering of the emperor and his empress and concubines was served in different areas. Besides, the "Thirteen Departments" encircled by earth mounds to the west were also auxiliary facilities. In the 13 courtyards were situated the Department of Four Imperial Affairs[2], Department of the Sympathi-que Clocks, Department of Boat Service, Department of Eunuchs and Maids[3], Department of Rites, Department of Medicine, Department of Fowling Guns, and Palace of Heavenly Purity. These departments covered all aspects of the daily life of the imperial family, engineering, ordnance and management.

[2] A department in charge of the emperor's clothing, weaponry and nighttime watch.

[3] A department in charge of the penalty of eunuchs and maids.

# 5.2 Landscape tours

**In this section, you will get an idea about:**
1. How frequently did the Qing monarchs travel between the capital city and the THFG?
2. How was Qianlong's life like as a "homebody" in Yuanming Yuan?
3. What were the activities for Emperor Qianlong when he went to Changchun Yuan and the Three Hills?

The everyday schedule of the emperor was documented in such detail as to cover where he went, what he did, whom he met, and even what he wore and what he ate, giving an impression that the emperor had no privacy at all. Statistics has revealed incredibly abundant daily activities for Emperor Qianlong. Every day, from morning till night, inside Yuanming Yuan the emperor was not always burying his head into the documents on his desk at the Hall of Diligence in State Affairs or holding ceremony at the Hall of Justness and Honesty and coming back to his bedchamber at the Peace over the Nine Prefectures for the night. Instead, he would often tour the individual scenic spots at the Three Hills, Changchun Yuan, and Yuanming Yuan by boat, on foot, on sedan, on horseback, or even by ice sled. Of course, this does not mean that the emperor was trying to be indolent. His trips to these places were usually made for definite purposes. To give a better explanation, we have sorted out his tours into the following three modes and visualized the routes to bring back the daily routine of the garden owner to the furthest extent.

## Mode 1: Traveling between the capital and Haidian

### By land (route 1):

Forbidden City—Xizhi Gate—Gaoliang Bridge—Haidian Town—Yuanming Yuan (e.g., on the 8th of the 1st month of the 21st year of Qianlong's reign)

This was the route by which the emperor went straight from the imperial palace in Beijing to Yuanming Yuan without a break on the way. It was also the most usual land route. As marked by the red and orange lines in the map: The emperor set out of town from the Imperial City to the Xizhi Gate, crossed Gaoliang Bridge and travelled northwestward for about 7 km by sedan, passed Haidian Town, took the imperial stone road east of Changchun Yuan and arrived at Yuanming Yuan.

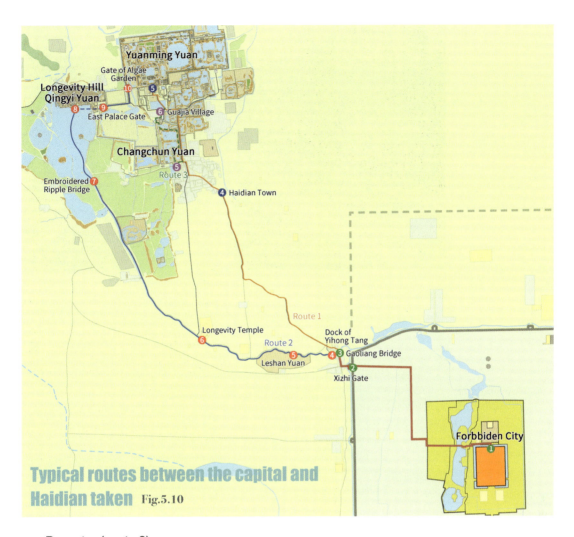

Typical routes between the capital and Haidian taken Fig.5.10

**By water (route 2):**

Forbidden City—West Palace—Xizhi Gate—Yihong Tang—Longevity Hill cum Qingyi Yuan—Yuanming Yuan (e.g., on the 25th of the 4th month of the 21st year of Qianlong's reign)

This was the route by which the emperor travelled from the imperial palace to Yuanming Yuan by way of Qingyi Yuan, when time was not so tight. It was also the most usual water route. As marked by the blue line in the map: The emperor set out of town, went northwestward, crossed the Gaoliang Bridge and started a boat cruise from the wharf. During the trip he might pass Leshan Yuan (Philanthropist's Garden), Temple of True Awakening and Longevity Temple, where he might stop for a short break. After arriving at the Guangyuan Sluice, as the water level was different, he needed to take another boat. After arriving at Qingyi Yuan across a distance of 8.5 km along the Long River (Fig.5.11), he entered the Kunming Lake from the inlet of the water way to its south (i.e., the Embroidered Ripple Bridge), then passed the Phoenix Mound, South Lake Island and Seventeen-Arch Bridge, landed at the Longevity Hill wharf, toured the sight on a sedan, left from the East Palace Gate, travelled on land to the Gate of Algae Garden at the southwestern corner of Yuanming Yuan and entered the garden.

Fig. 5.11
Scenery of the Long River

## Mode 2: Garden-based life inside Yuanming Yuan

**Typical route 1: On Lantern Festival**

Forbidden City—Grand Palace Gate—Hermit Eternal Spring's Fairyland—Peace over the Nine Prefectures—Garden of Shared Happiness–Dual Cranes Studio —Garden of Shared Happiness—Peace over the Nine Prefectures—High Mountains and Long Rivers—Crossed Pavilion—Garden of Shared Happiness—Peace over the Nine Prefectures (e.g., on the 13th of the 1st month of the 21st year of Qianlong's reign)

The time schedule of the first month was almost totally occupied by the prolonged Lantern Festival celebrations. It was the flourishing age and the national treasury was full of money. Celebrations once hit the historical high. The itinerary inside Yuanming Yuan was considerably different than the usual days. For example, Emperor Qianlong would spend this hard-earned spare time of the year staying in the Garden of Shared Happiness from breakfast time to dinner time, or traveling between the Peace over the Nine Prefectures, Garden of Shared Happiness, and High Mountains & Long Rivers for different activities. The emperor would have the empress dowager move from Changchun Yuan to the Hermit Eternal Spring's Fairyland at Yuanming Yuan. The mother and son stayed together for a longer time, too.

On that day (Fig.5.12-Fig.5.13), Emperor Qianlong, who had just arrived from the Imperial City, first went to see the empress dowager at the Hermit Eternal Spring's Fairyland. After a short break at the Peace over the Nine Prefectures, he went to the rear wharf to ride an ice sled to Garden of Shared Happiness for dinner. During the break after dinner and before the beginning of the lantern show in Garden of Shared Happiness, he would walk around the rarely visited Dual Cranes Studio (Broad Mind and Universal Justice) and a few other places. After the lantern show, he went back to the Peace over the Nine Prefectures for a brief pause. Then he hurried to the High Mountains & Long Rivers for the next activity: watching the wrestling and firework shows with the princes, dukes and ministers. Although it was already late, he took an ice sled at the wharf of the Crossed Pavilion at the Universal Peace & Harmony

Typical route 1: On Lantern Festival Fig. 5.12

again and traveled back to the Garden of Shared Happiness wharf to take an open sedan. When he returned to the Peace over the Nine Prefectures to sleep, it was already midnight. Obviously, the emperor was making a night trip.

**Typical route 2: Office life with occasional garden tours**

Peace over the Nine Prefectures—Embracing Faint Scent—Hall of Diligence in State Affairs—Goldfish Pond(i.e., Magnanimousness of Mind)—Peace over the Nine Prefectures—Blessing Sea Scenic Area—East Garden (i.e., Charngchun Yuan)—Garden of Shared Happiness—Peace over the Nine Prefectures (e.g., on the 23rd of the 6th month of the 21st year of Qianlong's reign)

During regular time when there was no major activity, Emperor Qianlong would have a rich day in Yuanming Yuan. In the morning hours, he would eat breakfast and handle state affairs at the Diligence in Government, Affection to the Virtuous. For all the rest of the day, he was quite relaxed and free, with pleasure activities held in various parts of Yuanming Yuan and Charngchun Yuan.

A busy day began from the morning hours (Fig.5.14). The eunuchs had already prepared

Fig. 5.13
Qing. Court Painter. The imperial ice sled in the *A Panorama of Empress Dowager Chongqing in Birthday Celebration* (Palace Museum)

165

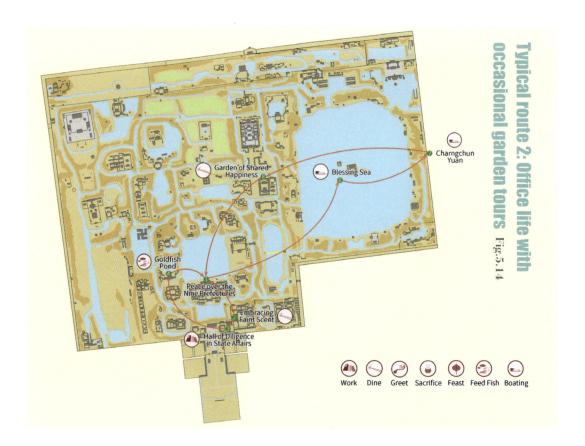

Fig. 5.14 Typical route 2: Office life with occasional garden tours

breakfast at the Embracing Faint Scent. After handling work and meeting with his ministers, Emperor Qianlong needed to change his clothes. Then he went to the Magnanimousness of Mind to feed the goldfish. After a short break back in the Peace over the Nine Prefectures, he set out on an unrestrained cruise. He took a boat from the rear wharf to Charngchun Yuan by way of the Blessing Sea scenic area. For this day, an additional fishing activity was also prepared. The emperor specifically wrote a poem on this activity. In the Blessing Sea scenic area, Immortal's Residence on Penglai Island, Taoist Wonderland on Fanghu Island, and Graceful & Peaceful Village (Concealed Beauty in the Fairy Caves) scenic spots were all frequently visited locations. Dinner was prepared at Garden of Shared Happiness. Sometimes he would also take a bath and change his clothes. In the leisurely hours after dinner, he would "stroll about" on a small boat and write a couple of poems. Finally, he went back to the Peace over the Nine Prefectures for the night. Sometimes on his way back, he would land on the Well-content Mansion to inspect the work of the painters. He might even give personal instructions.

### Typical route 3: Worship days on the 1st and 15th of the month

Peace over the Nine Prefectures—Mercy Cloud of Universal Blessing—Universal Peace and Harmony—Water-Moon Bodhimanda in Clouds—Buddha Tower—Sravasti City—Garden of Shared Happiness—Hall of Diligence in State Affairs—Prosperous Descendant Palace—Charngchun Yuan—Peace over the Nine Prefectures (e.g., on the 1st of the 7th month of the 21st year of Qianlong's reign)

For both the civilian world and the imperial court, the 1st and 15th of the month were important worship days. The Ghost Festival on the 15th of the 7th month and the Mid-Autumn Festival on the 15th of the 8th month were all the more important. Hence, this route shows how Yuanming Yuan served its worship function. On those occasions, in the morning hours, the emperor would rush among the various temples at Yuanming Yuan to worship the gods enshrined there. His working hours did not begin until all these activities were completed.

During these days (Fig.5.15), Emperor Qianlong first travelled to the Mercy Cloud of Universal Blessing on a boat to worship Buddha. Then he went to the wharf at the Universal Peace and Harmony. There he took a sedan to the Water-Moon Bodhimanda in Clouds, Buddha Tower (Vairocana's Magnificent Residence) and Sravasti City to worship Buddha. Then he had breakfast at the Garden of Shared Happiness. After breakfast, he took a boat to the Hall of Diligence in State Affairs to handle work and meet with his ministers. But the day's worship activities had not drawn to a close yet. He took a sedan to the Prosperous Descendant Palace on the southern bank of the Blessing Sea and the temples inside Charngchun Yuan to worship Buddha. Finally, he went back to the Peace over the Nine Prefectures for the night.

The number of worship activities in the garden throughout the year reached its peak on the Ghost Festival. This was a day for the Han people to worship their ancestors. It was also the Obon Festival for Buddhist believers. On that day, the emperor needed to worship Buddha at the imperial monasteries, including the Water-Moon Bodhimanda in Clouds, Vairocana's Magnificent Residence, Palace of Peace and Blessing, Sravasti City, and Prosperous Descendant Palace. At the same time, his work could not be left undone. After a busy day, the emperor watched the river lanterns on the water surface of Yuanming Yuan. That would be a sort of relaxation.

Typical route 3: Worship days on the 1st and 15th of the month Fig.5.15

## Mode 3: Tours from Yuanming Yuan to the Three Hills and Changchun Yuan

**Three-day tour to the Fragrant Hills (route 1):**

Yuanming Yuan—Jade Spring Hills cum Jingming Yuan—Fragrant Hills cum Jingyi Yuan (e.g., on the 14th of the 10th month of the 21st year of Qianlong's reign)

In the late autumn of Beijing, the sky was clear and the air was crisp. That was the day for Emperor Qianlong to leave for the Fragrant Hills to stay for a few days. The emperor was too busy to get such a chance to condition himself. That was the only once for him throughout the 21st year of his reign (1756).

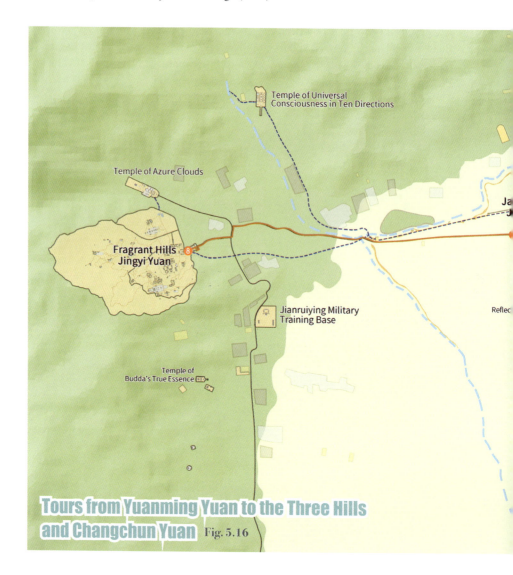

Tours from Yuanming Yuan to the Three Hills and Changchun Yuan Fig. 5.16

Early in the morning, the emperor set off on a four-men warm sedan from the Gate of Algae Garden of Yuanming Yuan. He passed the Archway of Graceful Scenery of Qingyi Yuan, travelled along the imperial road past the Town of Azure Dragon Bridge and Temple of Merits and Virtues, and then arrived at the Jade Spring Hills cum Jingming Yuan. After having breakfast there, he travelled on for another 4 km to arrive at the Fragrant Hills cum Jingyi Yuan by horse. According to poems made by the imperial order, during his stay at the Fragrant Hills, he visited the Green View Pavilion and Green Luxuriance Pavilion at Jingyi Yuan. He also visited the Temple of Azure Clouds and Temple of Buddha's True Essence nearby.

2 days later, on the 16th of the 10th month, after having breakfast in Jingyi Yuan, Emperor Qianlong decided to take a different route back home. First, he took an eight-men warm sedan to Jingming Yuan. There he toured a while and then left through the East Palace Gate. After passing the Picturesque Scenery Archway, he took a boat at the Tower at Lake Boundary (he might also leave through the South Palace Gate and take a boat cruise on the High-Level Water Lake or climb up the Tower of Reflection in Lake before

returning to the Tower at Lake Boundary). Next, he travelled to the Kunming Lake on the boat along the Jade River. After landing at the Longevity Hill, he viewed the sight there on sedan and returned to Yuanming Yuan. Seeing that there was still some surplus time, he took an ice sled to Charngchun Yuan and had dinner at the Classic Books Hall. On his way back, he passed the Graceful & Peaceful Village and finally returned to the Peace over the Nine Prefectures for the night. That marked the end of the full three-day tour.

### A short break at the Longevity Hill (route 2):

Yuanming Yuan—Longevity Hill cum Qingyi Yuan—Yuanming Yuan (e.g., 23rd of the 7th month of the 21st year of Qianlong's reign)

As we have mentioned in Chapter 4, to punish himself for breaking his word, Emperor Qianlong would normally visit Qingyi Yuan in the morning. Specifically, he might go there after having breakfast and handling work at the Diligence in Government & Affection to the Virtuous of Yuanming Yuan (as he did that day). He might also set out straight to Qingyi Yuan from his bedchamber, have breakfast and view the sight there, then return to Yuanming Yuan (as he did on the 9th of the 7th month). Apart from the Longevity Hill, the farther-away Taoist Temple of Immense Benevolence, Black Dragon Pool and Temple of Great Awakening were also destinations for the emperor after leaving Yuanming Yuan, mainly for consecration or worship purposes.

### A greeting to the empress dowager (route 3):

Yuanming Yuan—Changchun Yuan—West Garden—Longevity Hill cum Qingyi Yuan—Tower at Lake Boundary—Jade Spring Hills cum Jingming Yuan—Yuanming Yuan (e.g., on the 11th of the 7th month of the 21st year of Qianlong's reign)

This was a route by which the emperor fulfilled his filial duty to his mother, contemplating on state affairs and moulding his temperament. Emperor Qianlong started from the Gate of the Virtuous of Yuanming Yuan and took an eight-men open sedan into Changchun Yuan through the northwestern gate. After visiting his mother at the Gallery of Gathering Phoenix, he came to the Book House of Tracing the Origin in the neighboring West Garden to contemplate on governance strategies or recollect memories about his grandfather Emperor Kangxi. Sometimes he would meet with officials here, too. After changing his clothes, he took the sedan to the Longevity Hill. Sometimes he might not stay in Qingyi Yuan, but take a boat at the wharf and go straight to Jingming Yuan. After cruising Jingming Yuan, he might take the same route back to Qingyi Yuan and Yuanming Yuan. Sometimes he might go straight back to Yuanming Yuan by land.

The route may have been highly variable after the emperor's visit to his mother in Changchun Yuan. South of this garden lay large tracts of paddy fields, the Temple of Imperial Moralization (Shenghua Si) and Quanzong Miao were there, too. The emperor would occasionally come here to check how well the crops were growing or take part in the worship activities at Quanzong Miao. The usual transport means included horse and boat.

## Summary on the landscape tour modes

Looking back on the three main landscape tour modes, besides guaranteeing safety, the self-contained system of the THFG also offered great liberty for the work and life of the imperial family members. Compared with the palace compound back in the Imperial City, here it seemed to be a paradise. In his spare time off work, Emperor Qianlong, as monarch of the country, was finally able to free himself from the rigidly structured imperial palace. He was able to reside and work in Yuanming Yuan, to convey his emotional feelings and contemplate in the study at the West Garden, to gather with his family in Changchun Yuan, and to mould his temperament at the Three Hills (Fig.5.17).

Pitifully, reality was not as perfect as we imagine. The enormous imperial detached palaces had already "booked" a big budget for the subsequent terms of Imperial Household Department. It also further exacerbated the emperors' inseparation between state and family affairs. In other words, apart from handling state affairs, they had to decide on or even design the remodeling, maintenance or other chores of the various gardens. These family affairs greatly distracted their limited energy. After all, not all emperors were as energetic as Emperor Qianlong, who could not only complete all sorts of things whether important or trivial, but also spare a lot of time for touring scenes.

**Fig. 5.17**
Old photograph of the imperial boat at Yihe Yuan (Summer Palace)

**Calendar of the Qing imperial family in the THFG**

## 1st lunar month

### Birthday of the Jade Emperor
**9th:** Peace over the Nine Prefectures-Worship and burn incenses.

### Lantern Festival
**13th—19th:** Garden of Shared Happiness-Enjoy opera and lantern show.

**14th:** Hall of Honoring Three Selflessnesses-Bestow feasts on the emperor's sons, grandsons and princes.

**15th:** Peace and Blessing Palace-Worship and burn incenses.

Hall of Justness and Honesty-Bestow feasts on visiting vassals.

Yuanming Yuan & Charngchun Yuan-Worship the Buddha.

Peace over the Nine Prefectures-Offer up Yuanxiao.

**16th:** Hall of Justness and Honesty-Bestow feasts on scholars and ministers.

During the period: High Mountains & Long Rivers-Bestow feasts on ministers, visiting vassals and concubines; Watch wrestling and fireworks show.

Hall of Justness and Honesty-Set off Lantern hill (Aoshan). (Means set off fireworks for celebration at daytime)

### Worship
**23th:** Temple of Gratitude & Yearning and Temple of Gratitude & Blessing in Changchun Yuan- Emperors worshiped former emperors and empress dowagers here after the 42th year of Qianglong Period.

## 2nd lunar month

### Pray for rainfall
**During the month:** Black Dragon Pool/ Jade Spring Hills· Jingming Yuan-Emperors went in person or sent officials to worship the Dragon God of Devine and Plentiful Rainwater, and the Dragon God of Spreading Kindness and Merciful Blessing respectively.

### Flower Fairy Festival
**15th:** Temple of Gathered All Spring.

One of Han traditional festivals. To commemorate the birthday of flowers, civilians go spring outing and admire flowers. Emperors sent officials here to worship and burn incenses.

Note: All places are in Yuanming Yuan except Changchun Yuan, Jingming Yuan, Jingyi Yuan, Taoist Temple of Immense Benevolence and Black Dragon Pool, which are already marked.

## 3rd lunar month

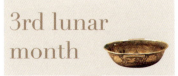

### Waterside Ritual Festival (Shangsi Festival)
**3rd:** Sitting on Rocks by Stream.

The Shangsi Festival is also called the Spring Bathing Day. Civilians have a custom of drifting cups on a meandering stream and go spring outing in the countryside that day. It can be deduced that Emperors commemorated the distinguished assembly at Orchid Pavilion here.

### Performing Plowing
**19th:** High Mountains and Long Rivers: Experience the hardship of farming.

### Pure Brightness Festival (Qingming Festival)
**2nd or 3rd lunar month:** Peace and Blessing Palace-Worship and burn incenses.

## 4th lunar month

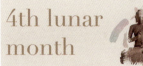

### Vesak Day
**8th:** Water-Moon Bodhimanda in Clouds and Sravasti City.

-The Vesak Day is also called Buddha's Birthday, which marks the birth of Sakyamuni. Vesak Fair would be hold at Sravasti City that day, and emperors also came to worship the Buddha.

### Birthday of Goddess of Bixia
**18th:** Prosperous Descendant Palace (Zither-like Water Sound over Two Lakes).

-Goddess of Bixia is the goddess of blessing all living and discerning good and evil in Taoism, and also an important faith of Northern China. Emperor Qianlong and the empress dowager once went here to worship, burn incenses and watch Temple Fair.

Taoist Temple of Immense Benevolence

-Both authorities and civilians worship and burn incenses here.

### Appreciate the beauty of flowers
**During the month:** Carving the Moon, Tailoring the Cloud and Graceful Scenery of the Western Peaks.

-Emperors went to the former place to appreciate peonies; to the latter one to appreciate Magnolia.

### Spring outing
**3rd or 4th lunar month:** Fragrant Hills· Jingyi Yuan.

-Emperors often set off from Yuanming Yuan to here for a temporary living, appreciating spring sceneries of the Three Hills.

## 5th lunar month

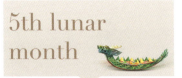

### Dragon Boat Festival
**5th:** Pavilion of Overlooking Yingzhou.

-Emperors went with empress dowagers, princes and ministers here to watch the dragon boat race beside the Blessing Sea.

### Day of Lord Guanyu Sharpening blade
**13th:** Vairocana's Magnificent Residence

-It is the day for Master Guanyu's seances. Emperors sent officials here to worship and burn incenses.

## 6th lunar month

### Appreciate lotus flowers
**During the month:** Lotus Fragrance (Spacious Paddies as Clouds).

-Emperors went with empress dowagers and princes here to appreciate lotus flowers.

## 7th lunar month

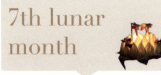

### Double Seventh Festival
**7th:** Graceful Scenery of the Western Peaks.

-Hold the banquet begging for cleverness, worship and burn incenses.

### Ghost Festival (Zhongyuan Festival)
Few days around the festival: Blessing Sea

-Go boating and watch river lanterns.

**15th:** Yuanming Yuan & Charngchun Yuan-Worship the Buddha.

Peace and Blessing Palace-Worship and burn incenses.

Temple of Gratitude and Yearning & Temple of Gratitude & Blessing at Changchun Yuan

-Worship and burn incenses.

# 8th lunar month

### Longevity Festival

**13th:** Garden of Shared Happiness.
-It was Emperor's Qianlong birthday. Envoys from vassal states came to celebrate when he was in Beijing, and feasts were held for celebration.

### Mid-Autumn Festival

**15th:** Mercy Cloud of Universal Blessing, etc.
-Worship the Buddha.
Garden of Shared Happiness, etc.
-Admire the moon view, enjoy the opera, eat moon cakes and play maze games.

# 9th lunar month

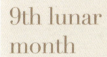 

### Double Ninth Festival

**9th:** Yuanming Yuan & Fragrant Hills·Jingyi Yuan
-Ascend a height and appreciate chrysanthemums.

### Banquets of Three Groups of Nine Elders

**9th:** Fragrant Hills·Jingyi Yuan.
-Qianlong Emperor bestowed the opportunity to visit gardens on 27 ministers in order to celebrate Empress Dowager Chongqing's 70th and 80th birthday.

# 10th lunar month

### Autumn outing

**9th or 10th lunar month:** Fragrant Hills· Jingyi Yuan.
-Emperors often set off from Yuanming Yuan to here for a temporary living, appreciating autumn sceneries of the Three Hills.

1st month-Watching lantern show | 2nd month-Admiring peach blossom | 3rd month-Drifting cups on a meandering stream

4th month-Going for spring outing | 5th month-Dragon boat racing | 6th month-Relaxing in the shade

7th month-Begging for cleverness | 8th month-Appreciating the moon | 9th month-Admiring the chrysanthemum

10th month-Getting self-portraits | 11th month-Practicing Zen meditation | 12th month-Enjoying the snow view

Note: This album is called the *Scroll Paintings of Twelve Lunar Months* by court painters of Qing dynasty (collected in the Palace Museum in Taipei).

# 5.3 Festival celebrations

**In this section, you will get an idea about:**
1.In what places were the state and family banquets held in the Qing dynasty?
2.What were the features of the breakfast, lunch and dinner menus on the Lantern Festival?

**Fig. 5.18**
Imperial clan banquet displayed at the Palace of Heavenly Purity of the Forbidden City

**Fig. 5.19**
Qing. Court Painter. Great Mercy and Eternal Blessing in the *Album of Paintings and Poems of Forty-Scenes of Yuanming Yuan*

**Fig. 5.20**
Interior of the Hall of Imperial Longevity at the Coal Hill

Like today, the Spring Festival was one of the most important festivity of the year. Besides, in the Qing dynasty, the emperor's birthday (Longevity Festival) and Winter Solstice were also regarded as two other important festivals. However, the venues for celebrating these festivals were not unchangeable. The Lantern Festival was the grandest imperial festivity with the most stable holding frequency in the THFG, or more precisely, Yuanming Yuan.

The Lantern Festival, together with Ghost Festival (the 15th of the 7th month) and Spirit Festival (the 15th of the 10th month), are collectively called "Three Yuan" ("Yuan" means beginning). These were the most important traditional festivals of the year. The formal date for the Lantern Festival was just the 15th of the 1st month. But here in Yuanming Yuan, the festivities would generally continue for around 7 days from the 13th till the 19th of the 1st month. During this time, the three-day state banquets and celebrations on the 14th, 15th, and 16th days of the month were the most important events. Besides the imperial family members, participants also included dukes and high ministers at court and envoys from vassal states. As there were too many people, the participants were separated into three lots according to their identity.

On the 14th of the 1st month, the imperial clan banquet was held at the Hall of Honoring Three Selflessnesses at the Peace over the Nine Prefectures. On this occasion, the emperor would gather with his wives, sons, grandsons, and brothers to celebrate the festival (Fig.5.18). On the 15th of the 1st month, at 12:00 and 13:45, the emperor would also eat meals here with his wives. A toasting ritual would be performed to the accompaniment of music.

On the 15th, the emperor was extremely busy the whole day. Besides attending banquets, a never-to-be-forgotten thing was to worship ancestors and gods. Hence, as soon as he got up, the emperor would worship ancestors at the Palace of Peace & Blessing, worship Buddha at the Vairocana's Magnificent Residence and Charngchun Yuan, and offer Yuanxiao (rice glue balls) to gods at the Peace over the Nine Prefectures. The main building of the Palace of Peace & Blessing was a huge, 9-bay palace. It measured 44.32 m across and 19.84 m deep, with a floor area of 879 m² (Fig.5.19) . Inside the shrines were the portraits of the ancestors, plaques singing the praise of them, and offerings such as food, books, and "Four Treasures of the Study (writing brush, ink stick, paper and inkslab)". Everything was arranged as if the spirits of the late emperors were still living there. From the rehabilitated Hall of Imperial Longevity at the Coal Hill (Fig.5.20), one can easily imagine how the Palace of Peace & Blessing looked at that time.

Besides restraining and complicating banquets, there were certainly relaxing and delighting activities, too. Garden of Shared Happiness (Fig.5.21) and High Mountains & Long Rivers would host a few evenings' boisterous festivities. At the Tower of Crystal Sound, the Grand Opera Tower at the Garden of Shared Happiness, besides the annual dramas, the diversity of color lanterns would make the palace very sumptuous. With a similar layout to the Tower of Unimpeded Sound inside the Forbidden City, the drama tower was comprised of a three-story stage and a performance tower behind it (Fig.5.22). The watching hall lay opposite the stage. On the sides of the courtyard were the seats of dukes and ministers. The construction of the opera tower was unusually intricate. The precise middle of the stage consisted of a square wellhead running from the roof down to the ground. Here, with the help of manually controlled mechanical apparatuses, all roles could "fly into the sky and fall into the ground" to present the various legends of immortals and Buddhas. We can see that the Qing people were not unprofessional at using machines. They simply used them in drama performance and European fountains.

If those lantern shows and drama performances at Garden of Shared Happiness were festivities typical of the Han nationality, here at the High Mountains & Long Rivers, the folk customs of the nomadic population were substantially highlighted (5.23). The Mongolian yurt that housed the emperor was huge in volume, with a diameter of approximately 23 m. The yurt was installed to face the south, looking toward the stage. On the large square carpet at the entrance, there would be thrilling wrestling performances. Overall, one would sense the striking features of nomads.

The main hall of the scenic area, the Tower of High Mountains & Long Rivers, was a two-story, nine-bay west-facing building. It measured 32.13 m across and 8.13 m deep. The hall was lined up with longevity lanterns on one side and Aoshan lanterns on the other. On the flat ground to the west were placed performance facilities like swings, copper ropes and firework shelves (Fig.5.24-Fig.5.25). Further west, the big square was the place for setting off fireworks. When the emperor and his wives came up here, the field of vision was very broad. Firework show was performed afterwards, in a variety and number far beyond people's imagination. In his poem, Emperor Daoguang so described, "**The shining lanterns were brighter than the stars and the moon. Layers upon layers of fireworks were as diverse as fish and dragons in water**".

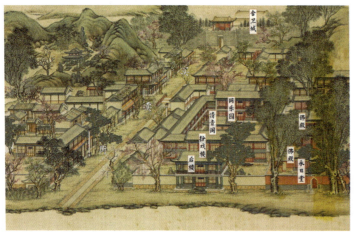
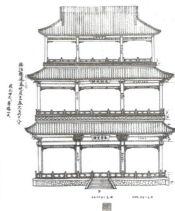

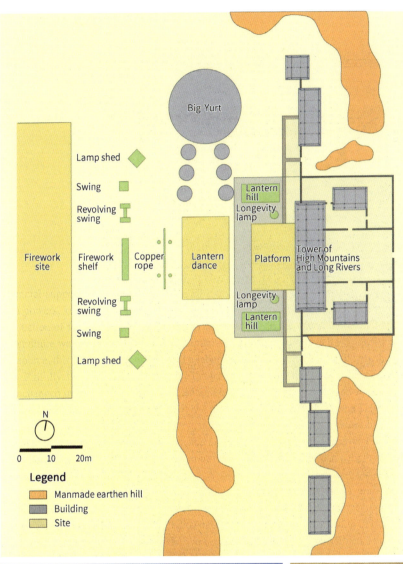

**Fig. 5.21**
Qing. Court Painter.
Garden of Shared
Happiness and Marketing
Street in the *Album of
Paintings and Poems of
Forty-Scenes
of Yuanming Yuan*

**Fig. 5.22**
Qing, Style House.
*Elevation View of the
Tower of Crystal Sound
at the Garden of Shared
Happiness*

**Fig. 5.23**
Restored plan of the
Lantern Festival Firework
Entertainment at the High
Mountains & Long Rivers
(Redrawn from the book
*Exploration of Gardening
Art of Yuanming Yuan*
by Jia Jun)

**Fig. 5.24**
The restored
longevity lanterns of the
Forbidden City

**Fig. 5.25**
Qing. Court Painter.
*Scroll Painting of Hongli's
Recreation in Lantern
Festival* (Palace Museum)

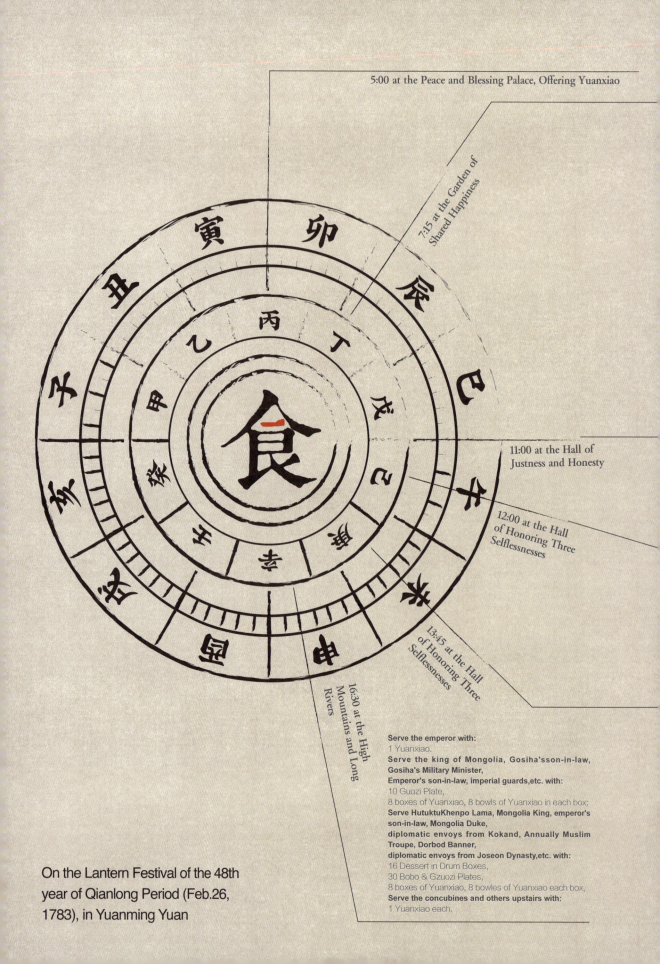

# IX

## Lantern Festival agendas and menus at Yuanming Yuan

**Serve the emperor with:**
1 Glutinous Millet,
1 Cubilose Hot Pot,
1 Cubilose and Shredded Fat Chicken,
1 Stewed Duck with Winter Bamboo Shoots and Duck Kidney,
1 Shredded Meat and Wild Rice Shoots,
1 Stewed Duck with Mushroom,
1 Duck Broth,
1 Hexigten Mongolia Specialty,
1 Steamed Fat Chicken,
1 Deer Tail Paste,
1 Chopped Pheasant,
1 Steamed Duck and Deer Tail Plate,
1 Simmered Pork Plate,
1 Deer Tail,
1 Dismember Whole Sheep,
1 Simmered Deer,
1 Simmered Pork,
1 Bamboo Shaped Steamed Bun,
1 Chicken and Sheep Tripe Dumpling,
1 Steamed Vegetable Bun.

**Serve the concubines with:**
1 hot pot,
1 Side Dish in Sunflower-shaped Enamel Box,
4 Side Dish in Enamel Plates,
1 Salted Meat,
1 Noodle with Cubilose and Three Delicacies,
4 tables of additional food,
18 Bobo,
8 Dry and Wet Desserts,
14 Milk,
2 vegetable dishes,
totally 2 tables of 40 dishes,
2 tables of 16 Meat Plates,
3 tables of Five Sorts of Mutton.

After breakfast, the emperor went to have Bobo Tables. Serve 10 Fruits in White Jade Plates for a table,
5 Bobo,
5 Guozi,
**along with:**
3 bowls, 3 plates and 2 boxes of hot Bobo,
2 bowls, 2 plates and 1 box of hot Guozi,
1 Yuanxiao.
**Serve 11 concubines with:**
6 Dessert in Drum Boxes,
1 Yuanxiao each;
**Serve 2 princesses with:**
2 Dessert in Drum Boxes,
1 Yuanxiao.
Serve Prince Rui, Prince Zhuang, Prince Xian, PrinceYi, Duke He, Duke Heng on eastside with:
3 Dessert in Drum Boxes.
Serve the officials of Military and Political Affair and imperial guards, etc. with:
5 Dessert in Drum Boxes.
Serve the Gosiha's son-in-law, etc. on westside with:
6 Dessert in Drum Boxes.

Serve 32 dishes in total:
4 Longlive Dry and Wet Dessert on both Sides,
1 Milk,
1 Milk Pancake,
1 Treble "Ten Thousand" Pickled Vegetable on both Sides,
1 Old Pickled Vegetable,
1 Mustard and Chinese Cabbage,
1 Soy Sauce.
On eastside, 1 table of 1st class dishes for Yu Fei, Chun Fei,
1 table of 2nd class dishes for the Tenth Princess and Xun Pin,
1 table of 3rd dishes for Lu Guiren and Bai Changzai,
on both Sides, 1 table of 1st class dishes for Ying Fei, Shun Fei,
1 table of 2nd class dishesf or Cheng Pin, Lin Guiren and Ming Guiren,
3 tables of Reddish Brown and White Vegetable in Bowl,
4 Dry and Wet Dessert,
4 Side Dish each table,
including 1 Soy Sauce.

Serve heated feast except soup, and invite the emperor arise from the throne and drink a toast,
After the emperor was seated and 4 chief managers invited banqueting tables in, the music ended,
Go out of the hall, and after the concubines were seated, serve the emperor's soup:
1 Cubilose and Duck Soup,
1 Polished Japonica Rice,
1 Cubilose and Shredded Chicken Soup,
1 Duck and Duck Kidney Soup;
Serve the concubines soup as soon as the box covers came out:
1 Polished Japonica Rice each,
1 Rice Noodle Soupwith Mutton and Poached Egg,
then serve Milktea as the emperor orders.
Turn to another feast and play music,
along with servinga table of wine and food, totally 32 dishes,
8 dual-paired Box Guozi,
8 triple-paired Box Dishes,
8 quadruple-paired Box Guozi.

**Serve the concubines with:**
wine feast, which had 15 dishes each table, totally 5 tables,
7 dishes and 8 Guozi,
then send 1 Yuanxiao;
Send the concubines and others Yuanxiao as soon as the box covers came out,
each one had 1 Yuanxiao,
after serving Yuanxiao, the music stopped, then send wine, and the music continued.
Chief Manager Xiao Yunpeng toasted a cup of liquor to the emperor,
After Xiao's kneeling to propose a toast, the emperor poured a cup of liquor as a reward,
Drink up, serve a cup of liquor and serve a cup of liquor to the emperor,
Go out of the hall, after serving a cup of liquor, serve everyone in the harem a cup of wine,
after sendingwine, the music stopped, serve Fruit Tea,
send the concubines Fruit Tea as soon as the bowl cover of Fruit Tea come out.

# 5.4 Military drill

**In this section, you will get an idea about:**
What were the respective functions and features of the 3 military systems in the THFG?

## Guard Divisions of the Eight Banners Forces of Yuanming Yuan

The garrisoning of military troops in the THFG region began in Kangxi period. In order to defend the safety of the imperial garden Changchun Yuan, the emperor would have a branch of the Eight Banners army from the capital stationed near the garden every day. The travel between the two places was pretty tiring. After Emperor Yongzheng succeeded to the throne, he used Yuanming Yuan as the detached palace in place of Changchun Yuan. In the 2nd year of Yongzheng's reign (1724), in imitation of the distribution regime of the Eight Banners army inside the capital city, the Guard Divisions of the Eight Banners

**Restored plan of the "Three Outer Division of the Eight Banners Forces"**
Fig. 5.26

Forces of Yuanming Yuan was stationed around Yuanming Yuan in the suburban area, to the north, east and south of the garden as standing troops according to the alignment of the left and right wings (Fig.5.26).

Within each of the banner barracks, the domiciles for high-rank officers all included 65 bays of rooms (except the large garrison of the Red-ringed White Banner); those for low-rank officers and ordinary soldiers all included around 1400 bays of rooms. In terms of the layout, all these banner barracks were surrounded by bounding walls, just like a small city compound. The rooms inside them were not only of equal sizes, but were aligned in a good order, contrasting strongly to the villages and hamlets around them.

If the barracks of the individual banners were places for the armymen to work, train and live, the Duipus ❺ around the bounding wall of the imperial gardens were the real lookout posts. The garrisoning of troops around the 4 imperial gardens of Yuanming Yuan, Charngchun Yuan, Changchun Yuan, and Qingyi Yuan during Qianlong period was recorded in historical texts. Needless to say, the largest number of troops were stationed around Yuanming Yuan, the garden of the greatest importance. The barracks numbered 779 men stationed at 76 watchhouses on the periphery of the bounding walls. By comparison, the barracks around Changchun Yuan, the garden where the empress dowager resided and Charngchun Yuan, the accessory garden to the east of Yuanming Yuan, numbered around 200 persons stationed at some 20 watchhouses. Although the bounding wall of Qingyi Yuan only contained the Longevity Hill (which covers almost 600,000 m$^2$), the emperor stationed 6 watchhouses and 61 persons. The defense was obviously much looser. However, this does not necessarily mean that the Longevity Hill and Kunming Lake could be accessed at will.

For Yuanming Yuan, the safety of the imperial family members was of paramount importance. Yuanming Yuan was enclosed by double stone walls. The garden was isolated from the outside world by a guarding river running outside the southern, western and northern walls. Only a few bridges were erected to connect the garden to the outside. The guard posts of the garrisoned troops were located right between two walls. To the east of the Grand Palace Gate were stationed Blue Banner Force, Red-ringed White Banner Force, White Banner Force, and Red-ringed Yellow Banner Force in counter-clockwise; to the west were stationed Red-ringed Blue Banner Force, White-ringed Red Banner Force, Red Banner Force, and Yellow Banner Force in clockwise, in strict accordance with the old rules. Obviously, this was the most heavily guarded imperial garden in the THFG region.

❺ Duipu, also pronounced as "Duibo", a Manchu term meaning the watchhouse where troops are stationed.

## Special Division of the Eight Banners Forces (Jianruiying)

In the 13th year of Qianlong's reign (1748), in order to pacify the revolts of the ethnic minorities in Greater Jinchuan and Lesser Jinchuan, Emperor Qianlong set up a special force at the Fragrant Hills, the Special Division of the Eight Banners Forces. There the soldiers were trained on a variety of skills, including aerial ladder, horseback archery, fowling piece, horse prancing, horse jumping, whip skill, sword skill, water fighting, boat sailing, and wind riding, and so on. Here the Eight Banners Forces were scattered around the Fragrant Hills cum Jingyi Yuan. There were in a total of 3,532 bays of barracks. Besides, 68 barbicans mimicking the buildings of ethnic minorities were also put

**Fig. 5.27** Jianruiying Military Training Base and one of the remaining barbicans

for training. The next year, the 11 m tall "Round City" was built here for the emperor to inspect the training (Fig.5.27).

## Outer Firearms Division of the Eight Banners Forces

In the 35th year of Qianlong's reign (1770), the emperor ordered to deploy the Outer Firearms Division (which echoed the Firearms Division back in the capital city) of the Eight Banners Forces on the western bank of the Long River near Qingyi Yuan for drilling the use of firearms equipment. There were two drill grounds. However, unlike the Guard Division and Special Division, a centralized layout was adopted. In total, it contained 1,024 bays of official rooms, 60 educational rooms, 6,038 cannon and armour rowhouses, and 3,176 gate towers in the surrounding bounding walls. Obviously, the place was a very enormous military base.

This marked the formal establishment of the "Three Outer Division of the Eight Banners Forces" military system.

### Emperor Qianlong's "Ten Military Achievements"

The grandiose Emperor Qianlong wrote in his article, "**I put down rebellions in Junggar twice, pacified Huibu once, fought Jinchuan twice, calmed down Taiwan once, defeated Burma and Annan once each, accepted the surrender of Gurkha twice, totaling ten.**" Although in 1860, the Qing army was unable to resist the warships and gunfires of the Anglo-French Allied Forces, in Qianlong period, the military strength of the Qing empire should have topped the world. His own pioneering in military construction also deserves recognition.

# 5.5 Administration and management

**In this section, you will get an idea about:**
1. Who were serving the imperial family in the THFG?
2. What farm crops were produced in the imperial gardens?
3. Why do we think the Yangshi Lei's Family represented the craftsmanship of the ancient Chinese people?

The enormous imperial palaces and multifarious affairs could never have operated right without a huge crew of organizers. After conquering the Central Plains, during Shunzhi's reign, the Qing dynasty set up an Imperial Household Department that took care of everything related to the imperial court. The head office of this department was called the "Chamberlain of the Imperial Household Department".

Under the Imperial Household Department were "Seven Departments and Three Courts". Among them, the institutions having the closest connection to the operation and management of the imperial gardens were the Court of Imperial Service and its subsidiaries.

The Court of Imperial Service was established in the 23rd year of Kangxi's reign (1684). As the name itself suggests, it was a department responsible for the management of the imperial palatial gardens. In the beginning, its responsibility was limited to the Three Seas of the West Palace, Coal Hill, South Palace and some smaller travel palaces, and Changchun Yuan, which was formally established in the western suburbs at a later time of the year. Over time, with the emergence of the THFG region, the Court of Imperial Service found itself dealing with increasingly more responsibilities and its divisions and workforce exploding, too. In the 53rd year of Kangxi's reign (1714), Jade Spring Hills Paddy Field Factory came into being. In the 3rd year of Yongzheng's reign (1725), as the rice produced in the crown land in and around Haidian Town and the Jade Spring Hills was far more than enough for the "**6 to 7 hundred dan (30000–35000 kg)**" needed to feed the imperial family, Emperor Yongzheng issued a decree that "**except that needed for growing grain for the imperial family, the rest farmland of the Paddy Field Factory shall be leased to the inhabitants nearby**". This meant that the paddy rice needed by the imperial family would just be produced on the imperial farmlands near the Temple of Merits and Virtues. This formally put the Paddy Field Factory under the management of the Imperial Household Department as a special function in charge of the paddy fields at the THFG.

In reality, apart from paddy fields, the capacity of the Court of Imperial Service also covered a variety of other lands, such as reed land, cattail land, lotus land and dry fields. The farm crops there, together with those growing inside the imperial gardens, not only

Tab.5.1 Court of Imperial Service's Management Function Allocation on Imperial Gardens

| Agency | Jurisdictional Limits | Remarks |
|---|---|---|
| Yuanming Yuan | Yuanming Yuan, Charngchun Yuan, Qichun Yuan, Xichun Yuan and Chunxi Yuan | Xichun Yuan and Chunxi Yuan separated successively from the management of Yuanming Yuan during Jiaqing and Daoguang's Period. |
| Changchun Yuan | Changchun Yuan, West Garden, Temple of Imperial Morolization, Quanzong Miao | |
| Qingyi Yuan (The Longevity Hill) | Qingyi Yuan, Temple of Merits and Virtues, watercourse near Jingming Yuan | The scope of the water surface and responsibilities include: "in addition to the **Azure Dragon Bridge sluice, from the north of Phoenix Mound, south of Azure Dragon Bridge, to the lake surface, river channel, embankment, bridge sluice, boat of Jingming Yuan, and the work of removing reed grass**" |
| Jingming Yuan (The Jade Spring Hills) | Jingming Yuan | |
| Jingyi Yuan (The Fragrant Hills) | Jingyi Yuan and surrounding imperial temples | Temples include Temple of Azure Clouds, Temple of Reclining Buddha etc. |
| Court of Imperial Service | Yihong Tang, Leshan Yuan and Longevity Temple, etc. The South Garden, paddy field, rivers, sluices, boats, etc. in the water system of the Long River, Golden River, Wanquan River outside the imperial gardens | The South Garden manage potted plants in Yuanming Yuan and other gardens |

supplied food for the imperial family, but also supplied funds for the maintenance and renovation works. Later, as Emperor Yongzheng was constantly settled in Yuanming Yuan, the affairs of the imperial gardens also became increasingly overloaded and important. As a result, all the large imperial gardens in the western suburbs were furnished with a separate management institution and a responsible official. Their administrative rank was roughly parallel to the Court of Imperial Service.

In Qianlong period, Changchun Yuan, Yuanming Yuan and the Three Hills were each furnished with a management department. Each of the departments managed a few smaller subordinate gardens. The work division was very clear, as shown in the Tab.5.1. However, although the three travel palaces at the Three Hills were independent themselves, they did not have sufficient autonomy regarding budgets. To some extent, they had to rely on the treasury of Yuanming Yuan. Hence, the most authoritative institutions should be the Court of Imperial Service and Yuanming Yuan.

## Staff of the management team and their incomes

As the imperial gardens encompassed a variety of contents in their composition, including hills and waters, palaces, plants, farm crops and animals, the activities of the imperial family were usually plentiful as described above. Keeping everything always in good order was therefore not an easy job. Hence a complete set of systems had to be in place. The emperor assigned different ranks of officials according to the area and importance of the individual gardens. They include different ranks of General Supervisor, Vice General Supervisor, Bithesi (person responsible for transcribing and translating Manchu texts), Chief of Warehouse Keeper, as well as employed Garden Resident, Chief of Garden Resident, Craftsmen, Sura (low-ranking service person), and Baitangga (unranked clerk). The number of persons was extremely large. The spending of the imperial family on manual labor alone was a big sum (Fig.5.28-Fig.5.29).

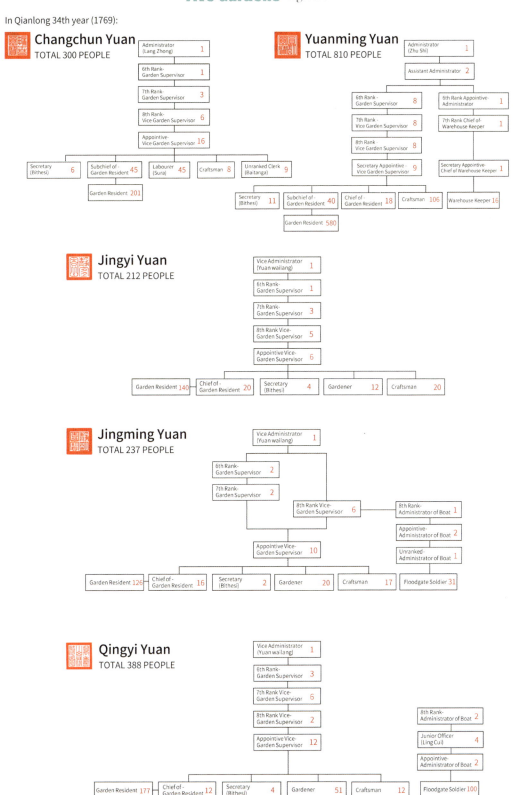

# Hierarchical structure of the management of the Five Gardens Fig.5.28

The eunuchs directly serving the imperial family members in the imperial court were also a very large group of people. In the 10th year of Jiaqing's reign (1805), as many as 620 eunuchs were working at Yuanming Yuan. Among them, the highest ranking was the 6th rank. For these eunuchs, the monthly income was 6 liang Silver and 6 hu [6] rice. Under this official were five grades. Among them the largest number fell upon the ordinary eunuchs. These eunuchs were also separated into 5 different subgrades, including eunuch skilled at using weapons, eunuchs working on board, ordinary eunuchs in and outside the gardens, and low-grade eunuchs inside the gardens. Their incomes ranged from 3 liang Silver and 4 hu rice down to only 2 liang Siver and 1.5 hu rice. Even so, the eunuchs were still earning much more money than the garden residents and craftsmen. Only their road to promotion was really too painstaking. And they had to give up a man's dignity, too.

## Plowing and Weaving

According to statistics of the 20th year of Jiaqing's reign (1815), at that time, the Paddy Field Factory was levying rents on 5,365,000 m² of paddy fields, 357,000 m² of dry fields, 968,000 m² of sandy land, 8,700 m² of cattail land, about 28,700 m² of homestead land and 58 bays of houses, amounting to 5,705.3545 liang Silver a year. This money, together with the money earned from selling grains, was used by the Court of Imperial Service for repairing the Three Seas of the West Palace and other imperial gardens. The majority of the paddy fields and dry fields were located in the THFG region.

Besides the plants on the rented lands named above, the THFG also possessed a disctinctive farm crop found in regions south of the Yangtze River—rape flower (Fig.5.30). In the 19th year of Qianlong's reign (1754), the emperor approved to grow rape flower from around Qingyi Yuan and on the banks of the Jade River. The rape flower fields on the banks of the Jade River extended as long as 3,200 m. The sea of golden flowers set off the Buddhist towers and pagodas at the Longevity Hill and Jade Spring Hills, rendering a magnificent sight. The ornamental and productive plant flourishes from March to May and bears fruit from the April to the June of each year. In the Qing dynasty, its seeds could yield 1.4 kg oil per 15 kg. The oil produced was delivered to the Three Official Granaries [7] for use.

In addition to the vast area outside the bounding walls, the imperial gardens themselves were also a "large farmstead". Besides the paddy fields in Changchun Yuan, Yuanming Yuan, and Jingming Yuan, fruit trees such as pear, apricot, peach, hawthorn, walnut and cherry trees, and so on were also planted at the Hill House of the Purple Jade Palace of Yuanming Yuan; vegetables and peach trees were grown at the Jade Spring Hills cum Jingming Yuan; fruit trees were also planted at the Fragrant Hills cum Jingyi Yuan. The large areas of lotus flower grown in the gardens were used not only for pleasure and producing lotus root, but also offerings to the temples and shrines like the Palace of Peace and Blessing. Everything was used to the fullest.

Besides farming, textile also constituted an important industry in the THFG. In

[6] Hu, a measure of rice in ancient China. In the Qing dynasty, one hu equals 50 liters and almost 50 kg of rice

[7] A food administrative agency under the Imperial Household Department

Fig. 5.29
Pie chart of the staff members of the Five Gardens

Fig. 5.30
Restored rape flowers and Rice of Western Beijing landscape at the Two-hills Park

the 16th year of Qianlong's reign (1751), Weaving and Dyeing Bureau at the Longevity Hill was founded along with Qingyi Yuan. It was located in the Pictures of Farming and Weaving scenic area northwest of the Kunming Lake. This building complex included spaces for all processes of production needed for textile: the weaving section at the front, reeling section at the back, dyeing section in the north, and silkworm house in the west. A worship activity was also held every year at the nearby Temple of Goddess of Silkworm. The layout of the Weaving and Dyeing Bureau was an intentional imitation of folk villages. Mulberry trees were also planted here and there for feeding silkworms. Obviously, in Qianlong's period, the design concept was already trying to achieve an integration between buildings and landscape, and between internal functions and external appearance.

## Source of funds for the imperial gardens

Before the late stage of the Qing dynasty, no evidence showed that the whopping sums spent on the imperial gardens were paid out of the national treasury. In reality, according to the proses made by the order to the emperor, the money used for the gardens all came from the Imperial Household Department, whether it be the construction of the Kunming Lake, the dredging of the Wanquan River; whether it be the work on the Quanzong Miao in Qianlong period, or for the Garden of Harmonious Interests project in Jiaqing period. In other words, the economy of the imperial gardens and the national treasury consisted of two separate systems. The "secret purse" of the imperial family was managed by the Imperial Household Department on behalf of the emperor. And in the past, it was totally impossible to disclose this mysterious account to the public. The money of the Imperial Household Department came from the proceeds from the imperial farmsteads and pawnshops, as well as incomes from the imperial family's usurious loans into the salt businesses. Many records indicate that the imperial money was circulated into the civilian world in forms of high interest rates and profit was gained from the businessmen. Seemingly, this "**helped improved the commercial power and contributed to monetary circulation**"(Letter to Emperor Qianlong from the Imperial Household Department). In the Qing dynasty, table salt was a monopolized industry. In order to afford the high taxes levied by the government and pay interests to the imperial family at the same time, salt dealers had no choice but to raise the salt price. Consequently, the ultimate victims were still the large population of civilians. A negative effect of this practice was the over-reliance of the imperial family on this source of income. When salt dealing began to fall, the enormous garden system up in Beijing found it hard to sustain any more. Moreover, as some researchers suggest, behind the clear discrimination of financial affairs between the inner and outer courts on the surface, the Imperial Household Department was actually eroding into the finance of the Ministry of Revenue and Population through its operation on salt administration. Hence, through the imperial gardens, we can gain insights into some of the rarely known financial "secrets" of the Qing dynasty.

## Imperial engineering works

The Qing imperial gardens was the property of the most powerful and wealthy man in the empire—the emperor. They were also enormous in size and exquisite in design. Hence a complete design and engineering system was definitely needed to serve them. The institution specifically set to undertake this mission was the "Style House (Yangshi Fang)" under the Imperial Household Department.

This institution had designed and constructed a huge number of imperial projects in and outside the capital city, including palace compounds such as the Forbidden City, the Three Seas of the West Palace, and the THFG; temples and altars such as the Imperial Ancestral Temple; imperial mausoleums such as the Western and Eastern Mausoleums of Qing Dynasty; city walls, such as the Zhengyang Gate. By "style", it means the drawings

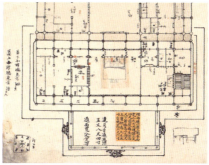

and models produced by craftsmen when designing a project. This is similar to what is done at a design institute for landscape architecture and architecture today. The design outputs would be presented to the emperor for review. They were finalized after repeated modifications and formally implemented after cost estimation.

Among the large team of craftsmen at the Style House was the very well-known "Yangshi Lei's" Family. The engineering drawings produced by them have been listed as a World Memory Heritage. Starting from Lei Fada in Kangxi period, all 8 generations of the Leis had been serving at the Style House. The most outstanding ones of the family were Lei Siqi and Lei Shengcheng. They worked as chief engineer for a long time during Kangxi, Yongzheng and Qianlong periods. It is fair to say that they were extraordinary technical talents and model workers in ancient China.

Today, from the Style House drawing archives, we can still see colored plan views, elevation views and even section views drawn with pigment for Chinese painting (Fig.5.31-Fig5.33). They bear neatly and reverently-written names and sizes of the buildings, their breadths and depths, and the ornaments, patterns and colors of the furniture inside. On the drawings, the earth mounds, rockeries and water systems in the gardens are integrated with the flexibly aligned building complexes. Few of the drawings also show the variety of plants and the distribution of minor garden items like sundials and brass vessels. In these artworks, the craftsmanship of the ancients is substantially demonstrated.

**Fig. 5.31**
Qing, Style House. *Master Plan of Yuanming Yuan of 1704#* (Palace Museum)

**Fig. 5.32**
Qing, Style House. *Plan of Tower of the Merged Colour of Waters and Skies at Yuanming Yuan* (National Library of China)

**Fig. 5.33**
Qing, Style House. Hot model of the Broad Mind and Universal Justice at Yuanming Yuan (Palace Museum)

# CONCLUSION

Nature has endowed the THFG with incomparable geographical advantages and scenic resources. The working people have relayed to develop more wonders in honor of these natural endowments. Men of letters have competed to praise it with poems and verses. These, together, impregnate the THFG with both natural and cultural glamour. The construction of the THFG since the Ming and Qing dynasties, especially in mid Qing period, pushed this beauty up to its historic pinnacle. Not only did it become an imperial political center outside the City of Beijing, it also occupies a pivotal position in the history of China and even the world with its immense size, complete functionality, superb artistic accomplishment, and profound cultural connotation.

Looking through the construction history of the THFG, we were surprised to discover that, in many a case, garden or landscape-making was not the foremost purpose, as exemplified by Qingyi Yuan, Jingming Yuan and Quanzong Miao. Instead, they were an artistic remodeling or beautification of natural environment based on water conservancy and agricultural development. Through a streamlined process of scientific siting, hills building and waters rearrangement, buildings layout, plants implanting, and culture implantation, the entire system was established. As the skill of this process matures, the aesthetic taste of the Chinese-style garden has also influenced the landscape design of other countries. Hence, it is never exaggerating to say that the THFG is a paradigm of the ancient Chinese art of building human settlements.

When we look back at the garden-based life of the Qing imperial family, on the one hand, they were deeply immersed in the Han culture. On the other hand, they also endeavored to merge their own ethnic features into the construction of the THFG. The emperors named the scenic spots with profound Confucianist, Buddhist, or Taoist significance. They held a diversity of activities in the gardens and created a copious number of poems and verses. They also indulged themselves deeply into the garden creation process... This lifestyle was totally irresistible compared with being barred behind the walls of the Forbidden City all day long. Isn't that the ideal life that has been sought after by literati for thousands of years? Although, in the feudal age, it was accessible for only the few upper-class minority, the underlying culture and art, especially the idea and spiritual pursuit of building a livable environment, are well worth to be shared by all mankind.

Neither can we forget the various craftsmen and managers that have left their sweat and blood in building and managing the gardens. Without a complete—and even harsh—management system, without their inventiveness and exquisite skills, without their attentive care and renovation, this massive system of the THFG would not have maintained a high artistic level and operated smoothly for almost two centuries. Abundant financial support, of course, was also indispensable. In the feudal age, the imperial family, by virtue of their own powers, kept pouring money into the creation and maintenance of gardens throughout the area, through the production of farm produce in the imperial farmsteads and the operation of the imperial capital nationwide. The expenditure was so huge that it is almost impossible to count how much was spent on the THFG at all. If we trace the source, this fortune was created by the laboring people. Hence it is justifiably the property of all people today.

Pitifully, the THFG has not remained in integrity through the years of history. Its ecological environment has also changed tremendously. Even the meaning of its name is rarely known to people today. Fortunately, the latest *Beijing Urban Master Plan (2016–2035)* has proposed wholly preserving the THFG and taking a series of actions. The significance of understanding the history of gardens is: Our ancestors have made countless painstaking attempts in seeking man-nature coexistence and meeting material and spiritual demands. Only by learning from the experiences they have gained throughout the thousands of years can people today preserve the character and feature of our own nation.

To this end, the authors of this book—a group of unimpressive students, set out on the expedition of research. Although we have yet to be fully fledged, we are not scared by the mountains of challenges in historical research. We never lose power because of the tediousness of work or incomprehension of other people. We try to find traces of history from seas of historical texts. We expand our focus from trivial details to the huge, entire region. Now we have preliminarily outlined how it looked before its destruction. In the meantime, in order for more people to know about and pay attention to the THFG, we have converted our academic result into more comprehensive contents. Through some popular promotion means, we have also organized a number of online and offline cultural activities that target at people of different professions and different ages. Our work has paid off with substantial social influence.

There is no end to research. There is no end to popularization. The THFG comes from the Jin dynasty more than 900 years ago and is going into the distant future. We hope that its landscape, culture, and art will deeply inscribed in our memories. We hope that it will become a paradigm area for inheriting and carrying forward the excellent historical culture and ecological civilization of the Chinese nation. We hope that its ecological benefits will continue far into the remote future... To realize these goals, we need to work together.

# GLOSSARY

## Names of Main Gardens

| English Name | Chinese Name | Paraphrase |
|---|---|---|
| Three Hills and Five Gardens (abbr. THFG) | 三山五园 | - |
| Longevity Hill | 万寿山 | - |
| Jade Spring Hills | 玉泉山 | - |
| Fragrant Hills | 香山 | - |
| Yuanming Yuan (Old Summer Palace) | 圆明园 | The Perfect and Wise Man's Garden |
| Charngchun Yuan | 长春园 | The Hermit Eternal Spring's Garden |
| Qichun Yuan | 绮春园 | Garden of Gorgeous Spring |
| Chunxi Yuan | 春熙院 | Yard of Spring and Prosperity |
| Xichun Yuan | 熙春园 | Garden of Prosperity and Spring |
| Changchun Yuan | 畅春园 | Garden of Smooth Spring |
| West Garden | 西花园 | - |
| Qingyi Yuan | 清漪园 | Garden of Clear Water and Ripples |
| Yihe Yuan (Summer Palace) | 颐和园 | Garden of Recuperating Harmony |
| Jingming Yuan | 静明园 | Garden of Tranquility and Brightness |
| Jingyi Yuan | 静宜园 | Garden of Pleasant Tranquility |
| Leshan Yuan | 乐善园 | Philanthropist's Garden |
| Purple Bamboo Yard | 紫竹院 | - |
| Quanzong Miao | 泉宗庙 | Temple of Springs' Origin |
| Shenghua Si | 圣化寺 | Temple of Imperial Moralization |
| Yihong Tang | 倚虹堂 | Hall of Leaning on Rainbow |
| Granted gardens | 赐园 | - |
| Jinchun Yuan | 近春园 | Garden of Approaching Spring |
| Tsinghua Yuan (Qing Dynasty) | 清华园（清） | Garden of Clear Water and Flourishing Woods |
| Minghe Yuan | 鸣鹤园 | Garden of Crane Whooping |
| Langrun Yuan | 朗润园 | Garden of Brightness and Moisture |
| Jingchun Yuan | 镜春园 | Garden of Mirroring Spring |
| Shuchun Yuan (Shihu Yuan) | 淑春园 （十笏园） | Garden of Mild Spring (Garden of Ten Ritual Tablets) |
| Weixiu Yuan | 蔚秀园 | Garden of Deep Luxuriance and Grace |
| Chengze Yuan | 承泽园 | Garden of Bearing Graciousness |
| Chenghuai Yuan (Garden of Imperial Academicians) | 澄怀园 （翰林花园） | Garden of Clearing Heart |
| Jixian Yuan (Hongya Yuan) | 集贤院 （宏雅园） | Yard of Gathering the Virtuous (Garden of Justness and Elegance) |
| Zide Yuan (Imperial Horses Stable) | 自得园 （御马园） | Garden of Enjoying Oneself |
| Prince Li's Garden | 礼王园 | - |
| Prince(Beizi) Zhi's Garden | 治贝子园 | - |

## Names of Scenes

| English Name | Chinese Name | Chinese Phoneticize |
|---|---|---|
| **Fragrant Hills cum Jingyi Yuan** | | |
| Twenty-eight Scenes of Jingyi Yuan | | |
| Hall of Diligence in State Affairs | 勤政殿 | Qinzheng Dian |
| Graceful Scenery Tower | 丽瞩楼 | Lizhu Lou |
| Boat House of Green Clouds | 绿云舫 | Lyuyun Fang |
| Modesty and Openness Studio | 虚朗斋 | Xulang Zhai |
| Jade-like Spring Cliff | 璎珞岩 | Yingluo Yan |
| Green Luxuriance Pavilion | 翠微亭 | Cuiwei Ting |
| Endless Green Woods | 青未了 | Qing Weiliao |
| Reindeer Hillside | 驯鹿坡 | Xunlu Po |
| Toad Rock | 蟾蜍峰 | Chanchu Feng |
| Tower of Dwelling in Clouds | 栖云楼 | Qiyun Lou |
| Pond of Knowing Happiness | 知乐濠 | Zhile Hao |
| Pines of Listening to Buddha Dharma | 听法松 | Tingfa Song |
| Fragrant Hills Temple | 香山寺 | Xiangshan Si |
| Green View Pavilion | 来青轩 | Laiqing Xuan |
| Xiabiao Steps | 霞标磴 | Xiabiao Deng |
| Crane Whooping Marsh | 唳霜皋 | Lishuang Gao |
| Fragrant Rock Chamber | 香岩室 | Xiangyan Shi |
| Jade and Milk Spring | 玉乳泉 | Yuru Quan |
| Gorgeous Autumn Woods | 绚秋林 | Xuanqiu Lin |
| Herbal Fragrance Mansion in Rain | 雨香馆 | Yuxiang Guan |
| Gorgeous Jade Peaks | 玉华岫 | Yuhua Xiu |
| Lotus Plain | 芙蓉坪 | Furong Ping |
| Bell Ring in Clouds | 隔云钟 | Geyun Zhong |
| Cascading Greenery Peaks | 重翠崦 | Chongcui Yan |
| Moon Dwelling Cliff | 栖月崖 | Qiyue Ya |
| Jade Ritual Tablet in Forest | 森玉笏 | Senyu Hu |
| Hill of Facing the Rising Sun | 晞阳阿 | Xiyang E |
| Fragrant Fog Cave | 香雾窟 | Xiangwu Ku |
| Other Scenes of Jingyi Yuan | | |
| Marketing Street | 买卖街 | Maimai Jie |
| East Palace Gate | 东宫门 | Donggongmen |
| Lofty Ideal Studio | 致远斋 | Zhiyuan Zhai |
| Ribbon-like Fountain and Screen-like Hills | 带水屏山 | Daishuipingshan |
| Watching Rising Clouds | 看云起 | Kanyunqi |
| Horologe Manufacture Office | 作钟处 | Zuozhongchu |
| Running Reindeer Garden | 奔鹿园 | Benlu Yuan |
| Dual Streams Spring | 双清泉 | Shuangqing Quan |
| Pramudita Garden | 欢喜园 | Huanxi Yuan |
| Deep in Bamboo Forest | 绿筠深处 | Lyuyunshenchu |
| Magnificent Brightness Temple | 洪光寺 | Hongguang Si |
| Pavilion under Green Canopy | 奋翠亭 | Yancui Ting |
| Eighteen Bends | 十八盘 | Shiba Pan |

| English Name | Chinese Name | Chinese Phoneticize | English Name | Chinese Name | Chinese Phoneticize |
|---|---|---|---|---|---|
| Heavy Clouds Pavilion | 多云亭 | Duoyun Ting | Hill in Floating Clouds | 飞云嶂 | Feiyun Wei |
| Tower of Telling Time | 知时亭 | Zhishi Ting | Gazebo of Rinsing Faraway Green | 漱远绿 | Shuyuan Lyu |
| Pavilion of Bringing Gracefulness | 致佳亭 | Zhijia Ting | Fragrant Clouds and Dharma Rain | 香云法雨 | Xiangyunfayu |
| Integrity and Peace | 正直和平 | Zhengzhiheping | Avatamsa Temple | 华严寺 | Huayan Si |
| Avalokitesvara Hall | 观音殿 | Guanyin Dian | Fragrant Rock Temple | 香岩寺 | Xiangyan Si |
| Autumn Pavilion | 有秋亭 | Youqiu Ting | Chamber in Clouds | 丛云室 | Congyun Shi |
| White Moonlight Pavilion | 约白亭 | Yuebai Ting | Courtyard of Attracting Cranes | 招鹤庭 | Zhaohe Ting |
| Mountain Mansion with Stepped Clouds | 梯云山馆 | Tiyunshanguan | Classic Books Hall | 含经堂 | Hanjing Tang |
| Hall of Cleaning Shoes | 洁素履 | Jie Sulyu | Calligraphy and Paintings Boat House | 书画舫 | Shuhua Fang |
| Western Hills Shimmering in Snow | 西山晴雪 | Xishanqingxue | Sumeru Temple | 妙高寺 | Miaogao Si |
| Pavilion of Lingering Moonlight | 延月亭 | Yanyue Ting | Hall of Containing Purity | 含醇堂 | Hanchun Tang |
| Temple as Magnificent as Jokhang | 宗镜大昭之庙 | Zongjingdazhao zhimiao | Pavilion in Lofty Clouds | 崇霭轩 | Chong'ai Xuan |
| | | | Hall of Praising Nature | 咏素堂 | Yongsu Tang |
| Tranquil Heart Studio | 见心斋 | Jianxin Zhai | Waterfall's Shadow Hall | 练影堂 | Lianying Tang |
| Pond of Feeding the Deer | 饮鹿池 | Yinlu Chi | Rapid Waterfall Belvedere | 飞淙阁 | Feicong Ge |
| Other Scenes of Jingyi Yuan | | | Pavilion of Waterfall-Hanging Eave | 挂瀑檐 | Guapu Yan |
| Buddha Theophany Hall | 圆灵应现殿 | Yuanlingyingxian Dian | Ripples Studio | 涵漪斋 | Hanyi Zhai |
| Tower of Fragrant Campaka Woods | 薝葡香林阁 | Zhanbuxianglin Ge | West Palace Gate | 西宫门 | Xigong Men |
| Broad Outlook Hall | 眼界宽 | Yanjiekuan | Benevolent Cultivation Palace | 仁育宫 | Renyu Gong |
| Tower of Placing Relaxation on Azure Glow | 青霞寄逸楼 | Qingxiajiyi Lou | Sacred Affinity Temple | 圣缘寺 | Shengyuan Si |
| Monk's Stave Spring | 卓锡泉 | Zhuoxi Quan | Moonlight Reflection Temple | 水月庵 | Shuiyue An |
| **Jade Spring Hills cum Jingming Yuan** | | | Tower of Encountering Pleasing Scenery | 赏遇楼 | Shangyu Lou |
| Sixteen Scenes of Jingming Yuan | | | Temple of Lotus Treasury World | 华藏海 | Huacanghai Si |
| Broadminded and Universal Justice | 廓然大公 | Kuorandagong | Water Gateway | 水城关 | Shuicheng Guan |
| Lotus-like Hill in Sunlight | 芙蓉晴照 | Furongqingzhao | Tower at Lake Boundary | 界湖楼 | Jiehu Lou |
| Gushing Spring of Jade Spring Hills | 玉泉趵突 | Yuquanbaotu | Other Scenes of Jingming Yuan | | |
| Hill Chamber with Bamboo Tea Stove | 竹炉山房 | Zhulushanfang | The First Spring Under Heaven | 天下第一泉 | Tianxiadiyi Quan |
| Comprehensive Imitation of Shengyin Temple | 圣因综绘 | Shengyinzonghui | Precious Pearl Spring | 宝珠泉 | Baozhu Quan |
| Poetic and Embroidered Cliff in Clouds | 绣壁诗态 | Xiubishitai | Gushing Jade Spring | 涌玉泉 | Yongyu Quan |
| Supervising Plowing by the Stream | 溪田课耕 | Xitiankegeng | Spring of Testing Ink | 试墨泉 | Shimo Quan |
| Cool Zazen Cave | 清凉禅窟 | Qingliangchanku | Spring of Tearing Silk Sound | 裂帛泉 | Liebo Quan |
| Winding Steps with Herbal Fragrance | 采香云径 | Caixiangyunjing | Spring of Splashing Pearls | 进珠泉 | Bengzhu Quan |
| Spring of Chinese Zither Sound in Canyon | 峡雪琴音 | Xiaxueqinyin | Dipankara Pagoda | 定光塔 | Dingguang Ta |
| Shadow of Pagoda at Jade Spring Hills | 玉峰塔影 | Yufengtaying | Sumeru Pagoda | 妙高塔 | Miaogao Ta |
| Refreshing Sound of Bamboo in Wind | 风篁清听 | Fenghuangqingting | Picturesque Scenery Archway | 湖山罨画牌坊 | Hushanyanhua Paifang |
| Mirror-like Water Reflecting the Sky | 镜影涵虚 | Jingyinghanxu | Tower of Reflection in Lake | 影湖楼 | Yinghu Lou |
| Scenery of Tearing Silk Sound Lake | 裂帛湖光 | Liebohuguang | **Changchun Yuan** | | |
| Distant Bell Ring in Clouds | 云外钟声 | Yunwaizhongsheng | Main Palace Gate | 大宫门 | Dagongmen |
| Fine Shade of Cloud-shaped Canopy | 翠云嘉荫 | Cuiyunhiayin | Hall of Nine Classics and Three Events | 九经三事殿 | Jiujingsanshi Dian |
| Other Scenes of Jingming Yuan | | | Secondary Palace Gate | 二宫门 | Er'gongmen |
| Peak Pavilion | 冠峰亭 | Guanfeng Ting | Hall of Kindness as Spring Sun | 春晖堂 | Chunhui Tang |
| Zhenwu Emperor Temple | 真武庙 | Zhenwu Miao | Hall of Day Lily with Longevity and Eternal Youth | 寿萱春永殿 | Shouxuanchunyong Dian |
| Dragon King Temple | 龙王庙 | Longwang Miao | Mansion at the Edge of Clouds | 云涯馆 | Yunya Guan |
| Luxuriance Mansion | 华滋馆 | Huazi Guan | Auspicious View Pavilion | 瑞景轩 | Ruijing Xuan |
| Studio of Identifying Heart | 甄心斋 | Zhenxin Zhai | Pavilion of Birds Flying and Fish Jumping | 鸢飞鱼跃亭 | Yuanfeiyuyue Ting |
| Green Clouds Hall | 翠云堂 | Cuiyun Tang | Peach Blossom Causeway | 桃花堤 | Taohua Di |
| The Gateway | 城关 | Chengguan | Fairy-herb and Orchid Causeway | 芝兰堤 | Zhilan Di |
| Deep in the Azure Clouds | 碧云深处 | Biyunshenchu | Lilac Causeway | 丁香堤 | Dingxiang Di |
| Crystal Sound Studio | 清音斋 | Qingyin Zhai | Vast and Hazy Pavilion | 苍然亭 | Cangran Ting |
| Splendor Hall | 含晖堂 | Hanhui Tang | Dragon King Temple | 龙王庙 | Longwang Miao |
| Pavilion of Sala Grove | 坚固林 | Jiangu Lin | Simplicity and Tranquility Hall | 澹宁居 | Danning Ju |
| Corridor of Chinese Zither Sound | 写琴廊 | Xieqin Lang | Main East Gate | 大东门 | Dadong Men |
| Winding Path Dividing the Water | 分鉴曲 | Fenjian Qu | Marketing Street | 买卖街 | Maimai Jie |
| Five-Arch Sluice | 五孔闸 | Wukong Zha | Studio of Appreciating Blossoms | 玩芳斋 | Wanfang Zhai |
| Extensive Green Hall | 延绿厅 | Yanlyu Ting | | | |

| English Name | Chinese Name | Chinese Phoneticize | English Name | Chinese Name | Chinese Phoneticize |
|---|---|---|---|---|---|
| Charming Pines Pavilion | 韵松轩 | Yunsong Xuan | Spacious Paddies as Clouds | 多稼如云 | Duojiaruyun |
| Dock | 船坞 | Chuanwu | Fish Jumping and Birds Flying | 鱼跃鸢飞 | Yuyueyuanfei |
| Studio of Working without Ease | 无逸斋 | Wuyi Zhai | North Distant Mountain Village | 北远山村 | Beiyuanshancun |
| Lotus Rock | 莲花岩 | Lianhua Yan | Graceful Scenery of the Western Peaks | 西峰秀色 | Xifengxiuse |
| Pine and Cypress Sluice | 松柏闸 | Songbai Zha | Book House Appropriate for Four Seasons | 四宜书屋 | Siyishuwu |
| Master Guanyu Temple | 关帝庙 | Guandi Miao | Taoist Wonderland on Fanghu Island | 方壶胜境 | Fanghushengjing |
| Bixiayuanjun Goddess Temple | 娘娘殿 | Niangniang Dian | Cleanse both Physically and Morally | 澡身浴德 | Zaoshenyude |
| Fragrant Blossoms Villa | 回芳墅 | Huifang Shu | Autumn Moon over the Calm Lake | 平湖秋月 | Pinghuqiuyue |
| Hall of Retaining Spring | 凝春堂 | Ningchun Tang | Immortal's Residence on Penglai Island | 蓬岛瑶台 | Pengdaoyaotai |
| Profound Literature Studio | 渊鉴斋 | Yuanjian Zhai | Hill House with Graceful Scenery | 接秀山房 | Jiexiushanfang |
| Fujun Temple | 府君庙 | Fujunmiao | Concealed Beauty in the Fairy Caves | 别有洞天 | Bieyoudongtian |
| Scattered Peaks | 疏峰 | Shu Feng | Zither-like Water Sound over Two Lakes | 夹镜鸣琴 | Jiajingmingqin |
| Trueness and Innocence | 太朴 | Taipu | Mirror-like Lake Reflecting the Sky | 涵虚朗鉴 | Hanxulangjian |
| Secondary East Gate | 小东门 | Xiaodong men | Broad Mind and Universal Justice | 廓然大公 | Kuorandagong |
| Book House Surrounded by Clear Stream | 清溪书屋 | Qingxi Shuwu | Sitting on Rocks by Stream | 坐石临流 | Zuoshinliu |
| Temple of Gratitude and Blessing | 恩佑寺 | Enyou Si | Distillery and Wind-blown Lotus | 曲院风荷 | Quyuanfenghe |
| Temple of Gratitude and Yearning | 恩慕寺 | Enmu Si | The Depths of Fairy Caves | 洞天深处 | Dongtianshenchu |
| Gazebo of Observing Billows | 观澜榭 | Guanlan Xie | **Other Scenes of Yuanming Yuan** | | |
| Pavilion of Gathering Phoenix | 集凤轩 | Jifeng Xuan | Hill House of the Purple Jade Palace | 紫碧山房 | Zibishanfang |
| Ruizhu Yard | 蕊珠院 | Ruizhu Yuan | Conform to the Nature of Plants | 顺木天 | Shunmutian |
| Main West Gate | 大西门 | Daximen | Imitated Sails Belvedere | 若帆之阁 | Ruofanzhige |
| Long Tower | 延楼 | Yanlou | Master Guanyu Temple | 关帝庙 | Guandi Miao |
| East Book House | 东书房 | Dongshufang | The Vast and Transparent Sky | 天宇空明 | Tianyukongming |
| Palace Gate | 宫门 | Gongmen | Algae Garden | 藻园 | Zao Yuan |
| Four Dwelling Palaces for Princes | 皇子四所 | Huangzisisuo | General Liumeng Temple | 刘猛将军庙 | Liumengjiangjun Miao |
| Book House of Tracing the Origin | 讨源书屋 | Taoyuan Shuwu | Temple of Gathered All Spring | 汇万总春之庙 | Huiwanzongchun zhimiao |
| Pavilion of Collecting Dew | 承露轩 | Chenglu Xuan | Pavilion of Literary Fountainhead | 文源阁 | Wenyuan Ge |
| Main North Gate | 大北门 | Dabeimen | Sravasti City | 舍卫城 | Sheweicheng |
| **Yuanming Yuan** | | | Marketing Street | 买卖街 | Maimai Jie |
| **Forty Scenes of Yuanming Yuan** | | | Garden of Shared Happiness | 同乐园 | Tongleyuan |
| Justness and Honesty | 正大光明 | Zhengdaguangming | Gate of the Virtuous | 出入贤良门 | Churuxianliang Men |
| Diligence in Government, Affection to the Virtuous | 勤政亲贤 | Qinzhengqinxian | Hall of Justness and Honesty | 正大光明殿 | Zhengdaguangming Dian |
| Peace over the Nine Prefectures | 九州清晏 | Jiuzhouqingyan | Hall of Diligence in State Affairs | 勤政殿 | Qinzheng Dian |
| Carving the Moon, Tailoring the Cloud | 镂月开云 | Louyuekaiyun | Nine Prefectures scenic area | 九州景区 | Jiuzhou Jingqu |
| Natural and Picturesque Scenery | 天然图画 | Tianrantuhua | Memorial Hall | 纪恩堂 | Ji'en Tang |
| Academy of Phoenix Trees | 碧桐书院 | Bitongshuyuan | Hall of Yuanming Yuan | 圆明园殿 | Yuanmingyuan Dian |
| Mercy Cloud of Universal Blessing | 慈云普护 | Ciyunpuhu | Hall of Honoring Three Selflessnesses | 奉三无私殿 | Fengsanwusi Dian |
| Merged Colour of Waters and Skies | 上下天光 | Shangxiatianguang | Hall of Peace over the Nine Prefectures | 九州清晏殿 | Jiuzhouqingyan Dian |
| Winehouse of Apricot Blossoms in Spring | 杏花春馆 | Xinghuachunguan | Blessing Sea | 福海 | Fu Hai |
| Magnanimousness of Mind | 坦坦荡荡 | Tantandangdang | Peace and Blessing Palace | 安佑宫 | Anyou Gong |
| Inclusiveness of Ancient and Modern | 茹古涵今 | Ruguhanjin | Profound Insight Hall | 洞明堂 | Dongming Tang |
| Hermit Eternal Spring's Fairyland | 长春仙馆 | Changchunxianguan | Congenital Grace | 前垂天贶 | Qianchuitiankuang |
| Universal Peace and Harmony | 万方安和 | Wanfanganhe | Culminant Prosperity | 中天景物 | Zhongtianjingwu |
| Spring Peach Blossoms of Wuling | 武陵春色 | Wulingchunse | Future Immortality | 后天不老 | Houtianbulao |
| High Mountains and Long Rivers | 山高水长 | Shangaoshuichang | Saint's Hall | 圣人堂 | Shengren Tang |
| Water-Moon Bodhimanda in Clouds | 月地云居 | Yuediyunju | Well-content Mansion (Imperial Academy of Calligraphy and Painting) | 如意馆 (皇家书画院) | Ruyi Guan |
| Great Mercy and Eternal Blessing | 鸿慈永祜 | Hongciyonggu | Gathering Spring in Heaven and Earth | 天地一家春 | Tiandi Yijiachun |
| Academy of Gathering Talents | 汇芳书院 | Huifangshuyuan | Hall of Cultivating Virtue | 慎德堂 | Shende Tang |
| Vairocana's Magnificent Residence | 日天琳宇 | Ritianlinyu | Studio of Imperturbation | 湛静斋 | Zhanjing Zhai |
| Simplicity and Tranquility in Mind | 澹泊宁静 | Danboningjing | Hall of the Foundation of Blessing | 基福堂 | Jifu Tang |
| Paddy Fragrance over the Water | 映水兰香 | Yingshuilanxiang | Self-Entertainment of Spring and Rock | 泉石自娱 | Quanshi Ziyu |
| Clear Water and Flourishing Woods | 水木明瑟 | Shuimumingse | Hall of Wistaria Shade and Blossoms | 藤影花丛殿 | Tengyinghuacong Dian |
| Mr. Lianxi's Land of Happiness | 濂溪乐处 | Lianxilechu | Stream of Knocking Jade | 鸣玉溪 | Mingyu Xi |

| English Name | Chinese Name | Chinese Phoneticize |
|---|---|---|
| Dual Cranes Studio | 双鹤斋 | Shuanghe Zhai |
| Crossed Pavilion | 十字亭 | Shizi Ting |
| Pure and Clean Place (Parisuddhi Place) | 清净地 | Qingjing Di |
| Embracing Faint Scent | 怀清芬 | Huai Qingfen |
| Graceful and Peaceful Village | 秀清村 | Xiuqing Cun |
| Prosperous Descendant Palace | 广育宫 | Guangyu Gong |
| Tower of Crystal Sound | 清音阁 | Qingyin Ge |
| **Charngchun Yuan** | | |
| Palace Gate | 宫门 | Gongmen |
| Stoicism Hall | 澹怀堂 | Danhuai Tang |
| Classic Books Hall | 含经堂 | Hanjing Tang |
| Mansion of Exquisitely Jade-like Rock | 玉玲珑馆 | Yu Linglong Guan |
| Studio of Reflection in Clear Water | 映清斋 | Yingqing Zhai |
| Garden of Taoist Fairyland | 小有天园 | Xiaoyoutianyuan |
| Studio of Considering Sustainability | 思永斋 | Siyong Zhai |
| Immortal's Mountain in the Sea and Broad Mind | 海岳开襟 | Haiyuekaijin |
| Islet of Diffusing Fragrance | 流香渚 | Liuxiang Zhu |
| Temple of Dharma's Wisdom | 法慧寺 | Fahui Si |
| Temple of Buddha's Solemn Appearance | 宝相寺 | Baoxiang Si |
| Waterside Orchids Hall | 泽兰堂 | Zelan Tang |
| Hall of Turning Boat Sail | 转湘帆 | Zhuanxiang Fan |
| Clustered Blossom Gazebo | 丛芳榭 | Congfang Xie |
| Lion Grove Garden | 狮子林 | Shizilin |
| Garden of Mirror-like Water | 鉴园 | Jian Yuan |
| Imitated Zhan Garden | 如园 | Ru Yuan |
| Garden of Flourishing Woods | 茜园 | Qian Yuan |
| **European Palaces** | | |
| Perspective Bridge | 线法桥 | Xianfa Qiao |
| Harmonious Wonder and Interest | 谐奇趣 | Xie Qiqu |
| Water Storage Tower | 蓄水楼 | Xushui Lou |
| Maze of Engraved Brick Wall | 万花阵 | Wanhua Zhen |
| Rare Birds Aviary | 养雀笼 | Yangque Long |
| Landscape of Fairyland | 方外观 | Fangwai Guan |
| Five Bamboo Pavilions | 五竹亭 | Wuzhu Ting |
| Hall of National Peace | 海晏堂 | Haiyan Tang |
| Throne of Appreciating the Fountains | 观水法 | Guanshuifa |
| Magnificent Fountains | 大水法 | Dashuifa |
| Distant and Oversea Landscapes | 远瀛观 | Yuanying Guan |
| Stone Archway | 石牌楼 | Shipai Lou |
| Perspective Hill | 线法山 | Xianfa Shan |
| Whorled Archway | 螺狮牌楼 | Luoshi Pailou |
| Perspective Paintings on the Walls | 线法画 | Xianfa Hua |
| **Longevity Hill cum Qingyi Yuan** | | |
| Graceful Scenery Archway | 涵虚罨秀牌楼 | Hanxuyanxiu Pailou |
| Hall of Diligence in State Affairs | 勤政殿 | Qinzheng Dian |
| Pavilion of Heralding Spring | 知春亭 | Zhichun Ting |
| Jade Billows Hall | 玉澜堂 | Yulan Tang |
| Pleasant Spring Hall | 怡春堂 | Yichun Tang |
| Hall of Happiness and Longevity | 乐寿堂 | Leshou Tang |
| Pavillion of Cultivating Clouds | 养云轩 | Yangyun Xuan |
| Gazebo of Facing Waterfowl | 对鸥舫 | Duiou Fang |
| Endless Charm Pavillion | 无尽意轩 | Wujinyi Xuan |
| Pavillion of Depicting Autumn | 写秋轩 | Xieqiu Xuan |
| Chairty and Blessing Tower | 慈福楼 | Cifu Lou |

| English Name | Chinese Name | Chinese Phoneticize |
|---|---|---|
| Temple of Immense Gratitude and Longevity | 大报恩延寿寺 | Da Baoenyanshou Si |
| Arhat Hall | 罗汉堂 | Luohan Tang |
| Revolving Archives | 转轮藏 | Zhuanlun Cang |
| Tower of Buddhist Incense | 佛香阁 | Foxiang Ge |
| Ratnamegha Bronze Pavilion | 宝云阁 | Baoyun Ge |
| Nest in Clouds and Pines | 云松巢 | Yunsong Chao |
| Fish and Algae Pavillion | 鱼藻轩 | Yuzao Xuan |
| Mansion of Listening to Orioles | 听鹂馆 | Tingli Guan |
| Strolling in Paintings | 画中游 | Huazhongyou |
| Pavilion of Saluting Rock | 石丈亭 | Shizhang Ting |
| Hall of Placing Emotions on Billows | 寄澜堂 | Jilan Tang |
| Marble Boat | 石舫 | Shifang |
| Five Saints Shrine | 五圣祠 | Wusheng Ci |
| Taoist Fairyland | 小有天 | Xiaoyoutian |
| Tower of Lingering Elegant Vision | 延清赏 | Yanqingshang |
| Wide Vision Studio | 旷观斋 | Kuangguan Zhai |
| Gateway of Cloud-Retaining Eaves | 宿云檐城关 | Suyunyan Chengguan |
| North Dock | 北船坞 | Bei Chuanwu |
| Watching the Clouds Rising | 看云起时 | Kanyunqishi |
| Gorgeous View Pavillion | 绮望轩 | Qiwang Xuan |
| Pavillion of Clear and Virid Water | 澄碧亭 | Chengbi Ting |
| Studio of Enjoying Leisure | 味闲斋 | Weixian Zhai |
| Garden of All-inclusive Spring | 赅春园 | Gaichun Yuan |
| Clouds Gathering Temple | 云会寺 | Yunhui Si |
| Sudarsana Temple | 善现寺 | Shanxian Si |
| Towering Pavillion | 构虚轩 | Gouxu Xuan |
| Suzhou Street | 苏州街 | Suzhou Jie |
| Nimbus Land on Sumeru Mountain | 须弥灵境 | Xumilingjing |
| Dragon King Temple | 龙王庙 | Longwang Miao |
| Blossom Clustering Belvedere | 花承阁 | Huacheng Ge |
| Picturesque Clouds Pavilion | 云绘轩 | Yunhui Xuan |
| Epiphyllum Belvedere | 昙花阁 | Tanhua Ge |
| Pavillion of Freshness after Snow | 霁清轩 | Jiqing Xuan |
| Garden of Harmonious Interests | 谐趣园 | Xiequ Yuan |
| Wenchang Emperor Belvedere | 文昌阁 | Wenchang Ge |
| Half-wall Bridge | 半壁桥 | Banbi Qiao |
| Lake Boundary Bridge | 界湖桥 | Jiehu Qiao |
| Waterside Dwelling | 水村居 | Shuicun Ju |
| Weaving and Dyeing Bureau | 织染局 | Zhiran Ju |
| Pictures of Farming and Weaving | 耕织图 | Gengzhi Tu |
| Binfeng Bridge | 豳风桥 | Binfeng Qiao |
| Sericulture and Ramie Bridge | 桑苎桥 | Sangzhu Qiao |
| Jade Belt Bridge | 玉带桥 | Yudai Qiao |
| Belvedere of Mirroring State Affairs | 治镜阁 | Zhijing Ge |
| Mirror Bridge | 镜桥 | Jing Qiao |
| Silk Bridge | 练桥 | Lian Qiao |
| Joyful Sight Hall | 畅观堂 | Changguan Tang |
| Hall of Appraising Talents | 藻鉴堂 | Zaojian Tang |
| Bright Scenery Tower | 景明楼 | Jingming Lou |
| Willow Bridge | 柳桥 | Liu Qiao |
| Phoenix Mound | 凤凰墩 | Fenghuang Dun |
| Embroidered Ripple Bridge | 绣漪桥 | Xiuyi Qiao |
| Spacious Pavilion | 廓如亭 | Kuoru Ting |
| Bronze Ox | 铜牛 | Tongniu |

| English Name | Chinese Name | Chinese Phoneticize | English Name | Chinese Name | Chinese Phoneticize |
|---|---|---|---|---|---|
| Seventeen-Arch Bridge | 十七孔桥 | Shiqikong Qiao | Pavilion of Happiness in Farming | 乐农轩 | Lenong Xuan |
| South Lake Island | 南湖岛 | Nanhu Qiao | Studio of Overlooking Faraway | 眺远斋 | Tiaoyuan Zhai |
| Hall of Overlooking Faraway | 鉴远堂 | Jianyuan Tang | Imperial Opera Department | 昇平署 | Shengping Shu |
| **Yihe Yuan (Summer Palace)** | | | Naval Academy | 水操学堂 | Shuicao Xuetang |
| **Partial Plan of the Longevity Hill Area** | | | Studio of Lingering Appreciation | 延赏斋 | Yanshang Zhai |
| Hall of Benevolence and Longevity | 仁寿殿 | Renshou Dian | **Nimbus Land on Sumeru Mountain** | | |
| Statement House | 奏事房 | Zoushi Fang | The Eastern Continent (Purvavideha) | 东胜身洲 | Dongshengshenzhou |
| Electric Light Office | 电灯公所 | Diandeng Gongsuo | The Western Continent (Aparagodaniya) | 西牛货洲 | Xiniuhuozhou |
| Yelyu Chucai Shrine | 耶律楚材祠 | Yelyu Chucai Ci | The Northern Continent (Uttarakuru) | 北俱芦洲 | Beijuluzhou |
| Gorgeous Sunset Tower | 夕佳楼 | Xijia Lou | The Southern Continent (Jambudvipa) | 南赡部洲 | Nanzhanbuzhou |
| Mansion of Book Collection | 宜芸馆 | Yiyun Guan | Deha | 提诃洲 | Tihezhou |
| Garden of Virtue and Harmony | 德和园 | Dehe Yuan | Videha | 毗提诃洲 | Pitihezhou |
| East Eight Offices | 东八所 | Dongbasuo | Sotha | 舍谛洲 | Shedizhou |
| Tea Serving House | 茶事房 | Chashi Fang | Uttaramantrina | 嗢怛罗漫怛里拿洲 | Wadaluo'mandali'nazhou |
| Gateway of Purple Cloud from the East | 紫气东来城关 | Ziqidonglai Chengguan | Camara | 遮末罗洲 | Zhemoluozhou |
| Gateway of Rosy Dawn in Chicheng Mountain | 赤城霞起城关 | Chichengxiaqi Chengguan | Varacamara | 筏罗遮末罗洲 | Faluozhemoluozhou |
| Natural Affinity of Water and Woods | 水木自亲 | Shuimuziqin | Kurdva | 矩拉婆洲 | Julapozhou |
| Promoting the Ethos of Benevolence | 扬仁风 | Yang Renfeng | Kaurava | 乔拉婆洲 | Qiaolapozhou |
| Pavilion of Containing Freshness | 含新亭 | Hanxin Ting | Sun Terrace | 日台 | Ritai |
| Pavilion in Woods and Clouds | 芸亭 | Yun Ting | Moon Terrace | 月台 | Yuetai |
| Gateway of Colorful Peaks | 千峰彩翠城关 | Qianfengcaicui Chengguan | Four Colour Stupas | 四色塔 | Sise Ta |
| Dense Green Pavilion | 重翠亭 | Chongcui Ting | White Stupa (Perfect Mirror-like Wisdom) | 白塔 | Bai Ta |
| Blessed Fortune Pavilion | 福荫轩 | Fuyin Xuan | Black Stupa (Wisdom of Universal Equality) | 黑塔 | Hei Ta |
| Relaxed Mood as Floating Clouds | 意迟云在 | Yichi Yunzai | Red Stupa (Wisdom of Perfect Conduct) | 红塔 | Hong Ta |
| Extended Longevity Hall | 介寿堂 | Jieshou Tang | Green Stupa (Wisdom of Profound Insight) | 绿塔 | Lyu Ta |
| Hall of Dispelling Clouds | 排云殿 | Paiyun Dian | **Forbbiden City** | | |
| Pavilion of Clear Water and Flourishing Woods | 清华轩 | Qinghua Xuan | Forbbiden City | 紫禁城 | Zijin Cheng |
| Shao Yong's Nest | 邵窝 | Shao Wo | Imperial Garden | 御花园 | Yuhua Yuan |
| Tower of Panoramic View of Lakes and Mountains | 山色湖光共一楼 | Shansehuguang gongyi Lou | Tower of Unimpeded Sound | 畅音阁 | Changyin Ge |
| Wealth and Longevity Never End | 贵寿无极 | Guishouwuji | Hall of Mental Cultivation | 养心殿 | Yangxin Dian |
| Natural Interest of Lakes and Mountains | 湖山真意 | Hushanzhenyi | Hall of Supreme Harmony | 太和殿 | Taihe Dian |
| Pavilion of Recepting Blessings | 承荫轩 | Chengyin Xuan | Hall of Preserving Harmony | 保和殿 | Baohe Dian |
| West Four Offices | 西四所 | Xisisuo | Palace of Gathered Elegance | 储秀宫 | Chuxiu Gong |
| Clear and Peaceful Boat (Marble Boat) | 清晏舫 | Qingyan Fang | Palace of Tranquil Longevity | 宁寿宫 | Ningshou Gong |
| Waterpoppy Bridge | 荇桥 | Xing Qiao | **Others Imperial Palaces** | | |
| Tower of Greeting the Sunrise | 迎旭楼 | Yingxu Lou | Imperial City | 大内 | Danei |
| Imitated Xiling | 小西泠 | Xiaoxiling | Coal Hill | 景山 | Jing Shan |
| Clearing Heart Belvedere | 澄怀阁 | Chenghuai Ge | Hall of Imperial Longevity | 寿皇殿 | Shouhuang Dian |
| Riverside Hall | 临河殿 | Linhe Dian | Western Mausoleums of Qing Dynasty | 清西陵 | Qingxi Ling |
| Hall of Rejuvenating Virtue | 德兴殿 | Dexing Dian | Eastern Mausoleums of Qing Dynasty | 清东陵 | Qingdong Ling |
| Ruyi Gate | 如意门 | Ruyi Men | South Palace | 南苑 | Nan Yuan |
| Gateway to the Misty Fairyland | 通云城关 | Tongyun Chengguan | West Palace | 西苑 | Xi Yuan |
| Temple of Buddha's Perfect Practice | 妙觉寺 | Miaojue Si | Three Seas (North, Middle, South Sea) | 三海 | Sanhai |
| Painted Blossoms Hall | 绘芳堂 | Huifang Tang | Garden of Abundant Rainfall | 丰泽园 | Fengze Yuan |
| Big Screen Wall | 大影壁 | Dayingbi | Imperial Mountain Summer Resort | 避暑山庄 | Bishu Shanzhuang |
| Gateway of Dawn Light | 寅辉城关 | Yinhui Chengguan | **Outside the gardens** | | |
| Glazed Pagoda | 琉璃塔 | Liuli Ta | **Hills** | | |
| Simplicity and Tranquility Hall | 澹宁堂 | Danning Tang | Red (Rock) Hill | 红山 (红石山) | Hong Shan (Hongshi Shan) |
| Utmost Blessing Belvedere | 景福阁 | Jingfu Ge | Shou'an Hills | 寿安山 | Shou'an Shan |
| Prolonged Longevity Hall | 益寿堂 | Yishou Tang | Wang'er Hills | 望儿山 | Wang'er Shan |
| Satisfaction Villa | 如意庄 | Ruyi Zhuang | Baiwang Hills | 百望山 | Baiwang Shan |
| | | | Lotus Leaf Hill | 荷叶山 | Heye Shan |
| | | | Incense Burner Peak | 香炉峰 | Xianglu Feng |

| English Name | Chinese Name | Chinese Phoneticize |
|---|---|---|
| Rivers and Lakes | | |
| Long River | 长河 | Chang He |
| Jade River | 玉河 | Yu He |
| South Dry River | 南旱河 | Nanhan He |
| North Dry River | 北旱河 | Beihan He |
| Qing River | 清河 | Qing He |
| Wanquan River | 万泉河 | Wanquan He |
| Lingjiao Pond | 菱茭泡子 | Lingjiao Paozi |
| Aqueducts | 引水石渠 | Yinshui Shiqu |
| Golden River | 金河 | Jin He |
| Fan Lakes | 扇面湖 | Shanmian Hu |
| High-Level Water Lake | 高水湖 | Gaoshui Hu |
| Water Storage Lake | 养水湖 | Yangshui Hu |
| Danling Pan | 丹棱沜 | Danling Pan |
| Cherry Ravin | 樱桃沟 | Yingtao Gou |
| Bridges and Sluices | | |
| Gaoliang Bridge | 高梁桥 | Gaoliang Qiao |
| Maizhuang Bridge | 麦庄桥 | Maizhuang Qiao |
| Baishi Bridge | 白石桥 | Baishi Qiao |
| Anhe Bridge | 安河桥 | An'he Qiao |
| Changchun Bridge | 长春桥 | Changchun Qiao |
| Azure Dragon Bridge | 青龙桥 | Qinglong Qiao |
| Guangyuan Sluice | 广源闸 | Guangyuan Zha |
| Baifu Weir | 白浮堰 | Baifu Yan |
| Towns and Villages | | |
| Town of Azure Dragon Bridge | 青龙桥镇 | Qinglongqiao Zhen |
| Haidian Town | 海淀镇 | Haidian Zhen |
| Bagou Village | 巴沟村 | Bagou Cun |
| Landianchang Village | 蓝靛厂 | Landian Chang |
| Chengfu Village | 成府村 | Chengfu Cun |
| Liulang Village | 六郎庄 | Liulang Zhuang |
| Guajia Village | 挂甲屯 | Guajia Tun |
| Wanquan Village | 万泉庄 | Wanquan Zhuang |
| Qing River Town | 清河镇 | Qinghe Zhen |
| Chuanying Village | 船营 | Chuan Ying |
| North Dock Village | 北坞 | Bei Wu |
| Middle Dock Village | 中坞 | Zhong Wu |
| South Dock Village | 南坞 | Nan Wu |
| Eight Water Yards and Temples on the Western Hills | | |
| Pond Water Yard | 潭水院 | Tanshui Yuan |
| Spring Water Yard | 泉水院 | Quanshui Yuan |
| Clear Water Yard | 清水院 | Qingshui Yuan |
| Fragrant Water Yard | 香水院 | Xiangshui Yuan |
| Dual Water Yard | 双水院 | Shuangshui Yuan |
| Devine Water Yard | 灵水院 | Lingshui Yuan |
| Sacred Water Yard | 圣水院 | Shengshui Yuan |
| Golden Water Yard | 金水院 | Jinshui Yuan |
| Temple of Great Awakening | 大觉寺 | Dajue Si |
| Temple of Cloud-like Dharma | 法云寺 | Fayun Si |
| Seclusion Temple | 栖隐寺 | Qiyin Si |
| Huangpu Temple | 黄普寺 | Huangpu Si |
| Temple of Dual Springs | 双泉寺 | Shuangquan Si |
| Temple of Golden Hill | 金山寺 | Jinshan Si |
| Eight Views of Yenching | | |
| Breeze at Taiye Lake in Autumn | 太液秋风 | Taiyeqiufeng |
| Luxuriance of Jade Island in Spring | 琼岛春阴 | Qiongdaochunyin |

| English Name | Chinese Name | Chinese Phoneticize |
|---|---|---|
| Golden Terrace in the Sunset | 金台夕照 | Jintaixizhao |
| Smoke Trees at the Ancient Jizhou City | 蓟门烟树 | Jimenyanshu |
| The Moon over the Lugou Bridge at Dawn | 卢沟晓月 | Lugouxiaoyue |
| Cascading Greenery of Juyong Pass | 居庸叠翠 | Juyongdiecui |
| Western Hills Shimmering in Snow | 西山晴雪 | Xishanqingxue |
| Gushing Spring of Jade Spring Hills | 玉泉趵突 | Yuquan Baotu |
| Military Layout of the Eight Banners Forces in THFG | | |
| Three Outer Divisions of the Eight Banners Forces | 外三营 | Waisan Ying |
| Guard Division of the Eight Banners Forces of Yuanming Yuan | 圆明园护军营 | Yuanmingyuan Hujunying |
| White-ringed Red Banner Force | 镶红旗 | Xianghong Qi |
| Red Banner Force | 正红旗 | Zhenghong Qi |
| Yellow Banner Force | 正黄旗 | Zhenghuang Qi |
| Red-ringed Yellow Banner Force | 镶黄旗 | Xianghuang Qi |
| White Banner Force | 正白旗 | Zhengbai Qi |
| Red-ringed White Banner Force | 镶白旗 | Xiangbai Qi |
| Blue Banner Force | 正蓝旗 | Zhenglan Qi |
| Red-ringed Blue Banner Force | 镶蓝旗 | Xianglan Qi |
| Elite Guard Division of the Eight Banners Forces | 精捷营 | Jingjie Ying |
| Outer Firearms Division of the Eight Banners Forces | 外火器营 | Waihuoqi Ying |
| Special Division of the Eight Banners Forces of the Fragrant Hills | 香山健锐营 | Xiangshan Jianruiying |
| Ethnic Minority Forces | 番子营 | Fanzi Ying |
| Imperial Racecourse | 马厂 | Ma Chang |
| Tower of Reviewing Troops | 阅武楼 | Yuewu Lou |
| Round City and Hall of Reviewing Troops(Jianruiying Military Training Base) | 团城演武厅 | Tuancheng Yanwuting |
| Main Temples and Mausoleums | | |
| Longevity Temple | 万寿寺 | Wanshou Si |
| Temple of True Awakening | 真觉寺 | Zhenjue Si |
| Temple of Great Wisdom | 大慧寺 | Dahui Si |
| Taoist Temple of Immense Benevolence | 广仁宫 | Guangren Gong |
| Temple of Merits and Virtues | 功德寺 | Gongde Si |
| Temple of Universal Consciousness in Ten Directions (Temple of Reclining Buddha) | 十方普觉寺 (卧佛寺) | Shifangpujue Si (Wofo Si) |
| Temple of Azure Clouds | 碧云寺 | Biyun Si |
| Temple of Immense Nurishment | 广润庙 | Guangrun Miao |
| Temple of Splendid Clouds | 妙云寺 | Miaoyun Si |
| Temple of Genuine Triumph | 实胜寺 | Shisheng Si |
| Temple of Buddhist Incense | 梵香寺 | Fanxiang Si |
| Temple of Buddha's True Essense | 宝谛寺 | Baodi Si |
| Temple of Buddha's Solemn Appearance | 宝相寺 | Baoxiang Si |
| Black Dragon Pool | 黑龙潭 | Heilong Tan |
| Emperor Jingtai's Mausoleum of Ming Dynasty | 明景泰陵 | Mingjingtai Ling |
| Temple of Emiting Luster | 遗光寺 | Yiguang Si |
| Temple of Storing Treasure | 宝藏寺 | Baozang Si |
| Tibetan Barbicans on the Western Hills | 西山大昭 | Xishan Dazhao |
| Temple of Merits and Virtues | 功德寺 | Gongde Si |
| Taoist Temple of Prosperous National Fate | 昌运宫 | Changyun Gong |
| Rectangular Tibetan Barbican | 方昭 | Fang Zhao |
| Round Tibetan Barbican | 圆昭 | Yuan Zhao |
| Temple of Great Awakening | 大觉寺 | Dajue Si |
| Temple of Awakening and Vitality (Grand Bell Temple) | 觉生寺 (大钟寺) | Juesheng Si (Dazhong Si) |

# Proper Noun

| English Name | Chinese Name | Chinese Phoneticize |
|---|---|---|
| **State Sectors and Imperial Departments** | | |
| hall | 堂、殿 | Tang/Dian |
| pavilion | 亭 | Ting |
| terrace | 台 | Tai |
| tower | 楼 | Lou |
| belvedere | 阁 | Ge |
| gazebo | 榭 | Xie |
| corridor | 廊 | Lang |
| pavilion | 轩 | Xuan |
| studio | 斋 | Zhai |
| house | 房 | Fang |
| boat house | 舫 | Fang |
| mansion | 馆 | Guan |
| pagoda | 塔 | Ta |
| chamber | 室 | Shi |
| **State Sectors and Imperial Departments** | | |
| Ministry of Personnel | 吏部 | Li Bu |
| Ministry of Revenue and Population | 户部 | Hu Bu |
| Ministry of Rites | 礼部 | Li Bu |
| Ministry of Military | 兵部 | Bing Bu |
| Ministry of Penalty | 刑部 | Xing Bu |
| Ministry of Engineering | 工部 | Gong Bu |
| Court of Supervision | 督察院 | Ducha Yuan |
| Court of Vassal State Affairs | 理藩院 | Lifan Yuan |
| Imperial Household Department | 内务府 | Neiwu Fu |
| Imperial College | 国子监 | Guozi Jian |
| Guanglusi Temple | 光禄寺 | Guanglu Si |
| Honglusi Temple | 鸿胪寺 | Honglu Si |
| Taipusi Temple | 太仆寺 | Taipu Si |
| Taichangsi Temple | 太常寺 | Taichang Si |
| Dalisi Temple | 大理寺 | Dali Si |
| Imperial Prince Department | 詹事府 | Zhanshi Fu |
| Memorial and Appeal Department | 通政司 | Tongzheng Si |
| Imperial Clan Department | 宗人府 | Zongren Fu |
| Imperial Astronomy Department | 钦天监 | Qintian Jian |
| Imperial Academy | 翰林院 | Hanlin Yuan |
| Duty House of Military and Political Affair Setup | 军机处值房 | Junjichu Zhifang |
| Duty House of Statement Department | 奏事处值房 | Zoushichu Zhifang |
| Duty house of the South Study | 南书房值房 | Nanshufang Zhifang |
| Court of Imperial Service | 奉宸苑 | Fengchen Yuan |
| **Personnel Titles** | | |
| Administrator | 郎中/主事 | Langzhong / Zhushi |
| Vice Administrator | 员外郎 | Yuanwailang |
| Garden Supervisor | 苑丞 | Yuancheng |
| Vice Garden Supervisor | 苑副 | Yuanfu |
| Appointive | 委署 | Weishu |
| Administrator of Boat Service | 催长 | Cui Zhang |

| English Name | Chinese Name | Chinese Phoneticize |
|---|---|---|
| Junior Officer | 领催 | Ling Cui |
| Bithesi | 笔帖式 | Bitieshi |
| Garden Resident | 园户 | Yuanhu |
| Chief of Garden Resident | 园户头目 | Yuanhu Toumu |
| Subchief of Garden Resident | 园隶 | Yuanli |
| Sura | 苏拉 | Sula |
| Baitangga | 效力柏唐阿 | Xiaoli Baitang'a |
| **Ancient Paintings and Calligraphys** | | |
| A Panorama of the Summer Palace and Surrounding Eight Banners Barracks | 《颐和园及附近八旗军营布局图》 | Yiheyuanjitujinba qijunyingbujutu |
| A Panorama of Emperor's Outing and Return of Worshiping Ancestor | 《出警入跸图》 | Chujingrubitu |
| A Panorama of Waterside Ritual at the Ladle Garden | 《勺园修禊图》 | Shaoyuanxiuxitu |
| A Panorama of the Fragrant Hills Palace | 《香山行宫图》 | Xiangshan xinggongtu |
| A Panorama of Jingyi Yuan | 《香山静宜园全貌图》 | Xiangshan jingyiyuanquanmaotu |
| Album of Paintings of Jingyi Yuan | 《静宜园图》 | Jingyiyuantu |
| A Panorama of 27 Elders Traveling in the Fragrant Hills | 《二十七老游香山图》 | Ershiqilao youxiangshantu |
| Album of Paintings of the Eight Views of Yenching | 《燕山八景图》 | Yanshanbajingtu |
| Note of the First Spring Under Heaven | 《天下第一泉记》 | Tianxiadiyiquanji |
| Album of Paintings and Poems of Forty Scenes of Yuanming Yuan | 《圆明园四十景图咏》 | Yuanmingyuan sishijingtuyong |
| Note of Yuanming Yuan | 《圆明园记》 | Yuanmingyuanji |
| Scroll Paintings of Forbidden Palaces in Twelve Months | 《十二禁禁图》 | Shierjinyutu |
| Copperplate Etching of European Palaces | 《西洋楼铜版画》 | Xiyanglou tongbanhua |
| Panorama of Ten-Scenes of the West Lake | 《西湖十景图卷》 | Xihushijing tujuan |
| A Panorama of Empress Dowager Chongqing in Birthday Celebration | 《崇庆皇太后万寿庆典图》 | Chongqing huangtaihouwanshou qingdiantu |
| Panorama of Water Conservancy in the Capital | 《都畿水利图》 | Dujishuilitu |
| Scroll Paintings of Twelve Lunar Months | 《十二月令图》 | Shi'eryuelingtu |
| **Festivals** | | |
| Flowers Festival | 花朝节 | Huazhaojie |
| Waterside Ritual Festival | 上巳节 | Shangsijie |
| Buddha's Birthday | 浴佛节 | Yufojie |
| feast at the Double Seventh Festival | 乞巧筵 | Qiqiaoyan |
| Emperor's Birthday /Longevity Festival | 万寿节 | Wanshoujie |
| The Meeting of Three Class of Twenty seven Elders | 三班九老会 | Sanbanjiulaohui |
| Lantern Festival | 上元节 | Shangyuanjie |
| Ghost Festival | 中元节 | Zhongyuanjie |
| Obon Festival | 盂兰盆节 | Yulanpenjie |

# REFERENCES

**1. Historical materials**

埃利松,2011. 翻译官手记[M]. 应远马,译. 上海: 中西书局.
北京市颐和园管理处, 2010. 清代皇帝咏万寿山清漪园风景诗[M]. 北京: 中国旅游出版社.
杜赫德, 2005. 耶稣会士中国书简集: 中国回忆录1-6[M]. 吕一民, 沈坚, 郑德弟, 译. 郑州: 大象出版社.
故宫博物院, 2000. 钦定总管内务府现行则例二种[M]. 海南: 海南出版社.
国家图书馆, 2016. 国家图书馆藏样式雷图档——圆明园卷初编[M]. 北京: 国家图书馆.
国家图书馆, 2016. 国家图书馆藏样式雷图档——圆明园卷续编[M]. 北京: 国家图书馆.
何瑜, 2014. 清代三山五园史事编年, 顺治—乾隆[M]. 北京: 中国大百科全书出版社.
何瑜, 2015. 清代三山五园史事编年, 嘉庆—宣统[M]. 北京: 中国大百科全书出版社.
刘侗, 于弈正, 1983. 帝京景物略[M]. 北京: 北京古籍出版社.
刘阳, 翁艺. 2017. 西洋镜下的三山五园[M]. 北京: 中国摄影出版社.
马戛尔尼, 巴罗, 2013. 马戛尔尼使团使华观感[M]. 何高济, 何毓宁, 译. 北京: 商务印书馆.
麦吉, 2011. 我们如何进入北京[M]. 叶红卫, 江先发, 译. 上海: 中西书局.
斯当东, 2016. 英使谒见乾隆纪实[M]. 钱丽, 译. 北京: 电子工业出版社.
王珍明, 张宝章, 易海云, 2006. 海淀文史·乾隆三山诗选[M]. 北京: 北京开明出版社.
吴蕾, 2017. 嘉庆圆明园静宜园诗[M]. 北京: 中国电影出版社.
香山公园管理处, 2008. 乾隆皇帝咏香山静宜园御制诗[M]. 北京: 中国工人出版社.
于敏中, 等, 1981. 日下旧闻考[M]. 北京: 北京古籍出版社.
中国第一历史档案馆, 1991. 圆明园[M]. 上海: 上海古籍出版社.
中华书局, 1986. 清实录·高宗纯皇帝实录. 北京: 中华书局.

**2. Modern monographs**

白日新, 2018. 圆明园盛世一百零八景图注[M]. 北京: 世界知识出版社.
北京通典图书有限公司, 2009. 颐和园手绘图（珍藏版）[M]. 北京: 五洲传播出版社.
北京通典图书有限公司, 2010. 圆明园盛景 [M]. 北京: 五洲传播出版社.
高大伟, 2011. 颐和园生态美营建解析[M]. 北京: 中国建筑工业出版社.
郭黛姮, 2009. 远逝的辉煌: 圆明园建筑园林研究与保护[M]. 上海: 上海科学技术出版社.
郭黛姮, 贺艳, 2010. 圆明园的"记忆遗产"——样式房图档[M]. 杭州: 浙江古籍出版社.
阚红柳, 2015. 畅春园研究[M]. 北京: 首都师范大学出版社.
何重义, 曾昭奋, 1995. 圆明园园林艺术[M]. 北京: 科学出版社.
侯仁之, 2009. 北京城的生命印记[M]. 北京: 三联书店出版社.
侯仁之, 2013. 北京历史地图集[M]. 北京: 文津出版社.
贾珺, 2009. 北京私家园林志[M]. 北京: 清华大学出版社.
贾珺, 2012. 北京颐和园（英文版）[Summer Palace] [M]. 北京: 清华大学出版社.
贾珺, 2013. 圆明园造园艺术探微[M]. 北京: 中国建筑工业出版社.
赖惠敏, 2016. 乾隆皇帝的荷包[M]. 北京: 中华书局.
李文君, 2017. 圆明园匾额楹联通解[M]. 北京: 故宫出版社.
刘阳, 2013. 谁收藏了圆明园[M]. 北京；金城出版社.
茅海建, 2014. 天朝的崩溃: 鸦片战争再研究[M]. 北京: 生活·读书·新知三联书店.
孟兆祯, 2014. 园衍[M]. 北京: 中国建筑工业出版社.
苗日新, 2010. 熙春园·清华园考: 清华园三百年记忆[M]. 北京: 清华大学出版社.
清华大学建筑学院, 2000. 颐和园[M]. 北京: 中国建筑工业出版社.
苏慈尔, 郝德士, 1962. 中英佛学辞典[M]. 台湾: 佛光出版社.
王其钧, 2007. 中国园林图解词典[M]. 北京: 机械工业出版社.
王珍明, 2000. 海淀古镇风物志略[M]. 北京: 学苑出版社.
吴祥艳, 宋顾薪, 刘悦, 2014. 圆明园植物景观复原图说[M]. 上海: 上海远东出版社.
武涛, 杨滨章, 2012. 风景园林专业英语[M]. 重庆: 重庆大学出版社.
夏成钢, 2008. 湖山品题·颐和园匾额楹联解读[M]. 北京: 中国建筑工业出版社.
徐卉风, 2015. 宫廷风·清帝南巡[M]. 上海: 上海远东出版社.

徐卉风, 2017. 宫廷风·乾隆与圆明园[M]. 上海: 上海远东出版社.
圆明园管理处, 2010. 圆明园百景图志[M]. 北京: 中国大百科全书出版社.
袁长平, 2018. 香山静宜园[M]. 北京: 北京出版社.
岳升阳, 夏正楷, 徐海鹏, 2009. 海淀文史·海淀古镇环境变迁[M]. 北京: 开明出版社.
张宝章, 2014. 三山五园新探[M]. 北京: 中国人民大学出版社.
张宝章, 2018. 玉泉山静明园[M]. 北京: 北京出版社.
张超, 2012. 家国天下——圆明园的景观、政治与文化[M]. 上海: 中西书局.
张德泽, 2001. 清代国家机关考略[M]. 北京: 学苑出版社.
张恩荫, 2003. 三山五园史略[M]. 北京: 同心出版社.
张天麟, 2010. 园林树木1600种[M]. 北京: 中国建筑工业出版社.
中国圆明园学会, 2007. 圆明园（全五册）[M]. 北京: 中国建筑工业出版社.
周维权, 2008. 中国古典园林史[M]. 北京: 清华大学出版社.
朱杰, 2000. 圆明园 [M]. 北京: 外文出版社.
CARROLL BROWN MALONE, 1966. History of the Peking Summer Palaces under the Ch'ing Dynasty[M]. New York: Paragon Book Reprint Corp.
CHENG LIYAO, 2012. Imperial Gardens[M]. 北京: 中国建筑工业出版社.
HU JIE, 2013. The Splendid Chinese Garden: Origins, Aesthetics and Architecture[M]. USA: Better Link Press.
VICTORIA M, CHA-TSU SIU, 2013. Gardens of a Chinese Emperor, Imperial Creations of the Qianlong Era, 1736—1796）[M]. Maryland: Lehigh University Press.
YONG-TSU WONG, 2016. A Paradise Lost: The Imperial Garden Yuanming Yuan[M]. 北京: 外语教学与研究出版社.

### 3. Journal articles

GAO SHAN, ZHU QIANG, ZHANG YIMING, 2018. When Time and Space Encounters: An Evolution Review of the Three Hills and Five Gardens in Beijing[J]. Beijing Planning Review, (9).
GUO CANCAN, ZHU QIANG, YIN HAO, et al., 2020。Study on the Plants and the Plant Landscape of Changchun Garden[J]. Chinese Landscape Architecture Magazine, 36(07).
LI ZHENG, LI XIONG, PEI XIN, 2015. Landscape Changes of the Paddy Fields in Western Beijing The Complexity and Contradiction in Its relationship with City[J]. Landscape Architecture Magazine, (12).
QUE ZHENQING, 2007. One More City Lost——The Royal Garden Cluster in the Northwestern Suburb of Beijing: The Three Hills and Five Gardens Declining in the Process of City Planning[J]. Zhuangshi Magazine, (11).
WANG YU, ZHU QIANG, LI XIONG, 2020. Research on the Plaques & Couplets and Artistic Conception of Changchun Garden[J]. Chinese Landscape Architecture Magazine, 36(06).
XIAO YAO, ZHU QIANG, ZHUO KANGFU, 2018. Plant Landscape and Management System of Beijing Royal Gardens, Qing Dynasty[J]. Landscape Architecture Magazine, (8).
YANG JING, LI JIANG, 2014. The Integral Visual Design of Imperial Gardens in the Western Suburb of Beijing[J]. Chinese Landscape Architecture Magazine, 30(02).
ZHU QIANG, 2018. Yuanming Yuan and the Spiritual World in Imperial Gardens[J]. Beijing Observation Magazine, (9).
ZHU QIANG, LI DONGCHEN, GUO CANCAN, et al., 2019. A Study on the Restoration and Gardening Method of Changchun Garden in Qing Dynasty[J]. Landscape Architecture Magazine, (2).
ZHU QIANG, LI XIONG, 2017. The Old Summer Palace and the Spiritual World in Imperial Gardens[J]. ICH Courier, (31).
ZHU QIANG, ZHANG YUNLU, LI XIONG, 2017. New Thoughts on "Three Hills and Five Gardens" Region in Beijing[J]. Journal of Chinese Urban Forestry, (1).

### 4. Dissertations

XU TONG, 2016. Study of Emperor Qianlong and the Image of the Western Hills of Beijing[D]. Beijing:Central Academy of Fine Arts.
YANG SHAN, 2017. A Study on the English Translation of Architecture Names in the Palace Museum[D]. Beijing:Beijing Foreign Studies University.
YIN LIANG, 2003. Research on Jingyi Garden and Jingming Garden in Qing Dynasty[D]. Tianjin : Tianjin University.
ZHANG DONGDONG, 2016. A Study on the Layout and Selective Parts of the Garden of Clear Ripples[D]. Beijing:Beijing Forestry University.
ZHANG LONG, 2006. The Analyse of Landscape Pattern & the Research of Architectural Layout in Qingyi Garden during Qianlong Reign[D]. Tianjin : Tianjin University.

# AFTERWORD

At the time this English version book is going to be published, my heart is filled with many emotional thoughts. Writing a book was not as easy as I had thought at first. It was a test on my professional and writing ability, as well as a trial on my willpower. In fact, I find myself grown a lot after this experience. As this book shoulders the responsibility of popularizing the history and culture of the Chinese imperial gardens to the public, especially to the foreign readers, the importance of its preciseness and readability is self-evident. For this reason, we have invested great efforts in verifying historical texts and organizing graphic expressions. We have also spent a lot of time trying to make details more accurate, in an effort to provide readers with an authentic history and wonderful reading experience. We were prepared to stay intimate with difficulties the moment we decided to have this done.

Throughout the process, we have been lucky enough to receive a lot of help from leaders, experts, and friends. I'm extremely grateful to Mr. Ye Liangqing and Ms. Gao Jie from the Culture Development and Promotion Center of Haidian District and Professor Zhang Baoxiu from Beijing Union University for their kind support for this book and our team, to China Forestry Publishing House for publishing this book, to Editor Ms. Sun Yao and the designer Mr. Liu Linchuan for their industrious work, and to Ms. She Sha for translating this book in vivid, accurate words. I'm indebted to my doctoral supervisor, Academician Meng Zhaozhen for giving us advice and encouraging us in his autograph, to Mr. Zhang Baozhang for giving us advice and providing valuable advice on how to improve our manuscript, and to Party Secretary Wang Hongyuan, President An Lizhe of Beijing Forestry University for being concerned about us all the time. I thank my master supervisor, Professor Li Xiong, for his kind instructions. I thank all leaders and teachers at the School of Landscape Architecture for their relentless support. I thank all reviewers for their valuable comments and opinions. I would like to thank associate professor Qian Yun for his guidance and help, as well as the development platform and support provided by the Beijing Siming Town Preservation Cooperation Program for these years.Many thanks to my dear team members for working side by side with me, and to previous members for offering unselfish support. Many thanks to Mr. Wu Xiaoping for providing those excellent photographs, to Ms. Xu Ming for valuable advices and encouragements. Besides, I also want to thank my family and friends for their constant concern and support.

Since it was founded in 2015, the Research Team of the Three Hills and Five Gardens of Beijing Forestry University has embraced a group of young members who are enthusiastic about traditional history and culture. In our research and popularization on the THFG, we have touched upon the pulse of history and harvested knowledge and pleasure. To me, that is the greatest delight of all. Throughout the compilation of this book, from planning, writing through to modification and correction, I felt the persistence of my colleagues and their endeavor to always make the best. I'm very proud of this team.

At last, this book of collective wisdom is going to be officially published to the world. Words cannot describe how excited I am now. I know our research on the THFG has just taken the first step, and there are a lot more historical puzzles ahead awaiting us to find the answer. I hope I will continue to work with my comrades and do our bit for the protection and inheritance of historical and cultural heritages.

**Zhu Qiang**
July, 2020

# A Dialog with the Team

The Research Team of the Three Hills and Five Gardens of Beijing Forestry University, a vibrant, young and voluntary group of garden theory explorers, with their diligence and readiness for innovation, set a challenging but very meaningful goal for these heritages: to launch a restoration research and popularization campaign on the THFG with professional knowledge. With their perseverence of publishing academic articles and books, filming documentaries, broadcasting programmes, holding exhibitions and public activities, making speeches for these 5 years, the team has gain a certain extent of influence of the public and the authority. They hope to do their best to make contribution to the revitalization of Chinese traditional culture and global culture communications. The team's Wechat Public Accout ID is "BFU_3Hills-5Gardens" and E-mail address is sswy_researchteam@163.com.

### Cong Xin

With love for the Chinese classic Landscape Architecture and curiosity about the lifestyle of the ancients, I joined my teammates in the research of the THFG. During the compilation of this book, I can feel the home-country-world sentiments and the vicissitudes of history contained in this group of classic gardens all the time. How to convey the true meanings of the Chinese text without losing its artistic charm is the destination we have been trying to arrive at. I hope this book will spread the inventive craftsmanship and vivid stories of the THFG farther into the other parts of the world.

### Wang Yize

Growing up in Beijing, I have a deep involvement with the THFG. Becoming a new member of this team offered a good opportunity to know and study more about the architecture of the imperial gardens. Just doing some fundamental work has already benefited me a lot, since it both enhances my academic ability and gains me deeper insight on the city where I live. I marvel at the integrity and systematicness of the THFG region. I hope the publication of this book will give foreign readers an idea about the magnificence of the traditional culture of the THFG and even the whole China.

### Song Qi

My first acquaintance with the THFG should have taken place in one of the art history classes in middle school, when I was attracted by its profound historic culture. Later on, as my knowledge deepens, I have come to realize that the THFG does not simply represent the extraordinary achievements in the Chinese classic garden, it also covers many other aspects like landscaping pattern, military system, and agricultural distribution. So, studying the history and evolution of the THFG is of great significance to the protection and development of the historic cultural heritages in this area. I'm very happy to be part of this research effort.

### Fu Jing

I'm very happy to join this team of dreams and passions and to contribute to the exploration and restoration of the THFG with my teammates. The deeper I'm involved in the compilation of this book, the more I feel the glamour of this cultural wealth. Behind the THFG is a colossal cultural system of a historic era, awaiting us to mine out its profound deposit and make it glow with a new life. Hope I will go farther along this road with my affection for the THFG.

## Cao Shuyi

I used to suppose that this cultural treasure was an inaccessible existence. As my knowledge deepens, I have come to see the excellence of its design and the profoundness of its history. Everything about it has grown familiar and intimate. During the research I felt the preciseness of academic verification as well as the pleasure of exploring history by myself. Participating in compiling this book is an unforgettable experience in my life. It also marks the beginning of my story with the THFG. I hope I will continue to contribute to the THFG together with the other members of the team.

## Du Yican

During my acquaintance with the THFG, reading literature gives me an opportunity of knowing some rarely-known historic details, such as the menu of Emperor Qianlong, the management system of the THFG, and the plowing and weaving procedure in the ancient times. The THFG is no longer a matter of gardens and landscapes, but a cross-disciplinary, multilevel, three-dimensional complex. Participating in this work is extremely meaningful to me. I hope this conglomeration of our painstaking effort will help readers know the multiple aspects of the THFG and contribute to the communication of the historic culture of the Chinese classic garden.

## Guo Jia

Participation in compiling this English version of the book was a tense, stimulating, yet fruitful challenge, since it was my first time to dig into the meanings of the imperial gardens that I have known for all these years. In the past, I never asked myself where those old-fashioned, artistic names originate, nor did I know that they could carry so many expectations, wishes, and aspirations. I hope our translation will help more people know the origins and implications of these gardens., Whether they belong to the past or still exist in reality, they should never be forgotten.

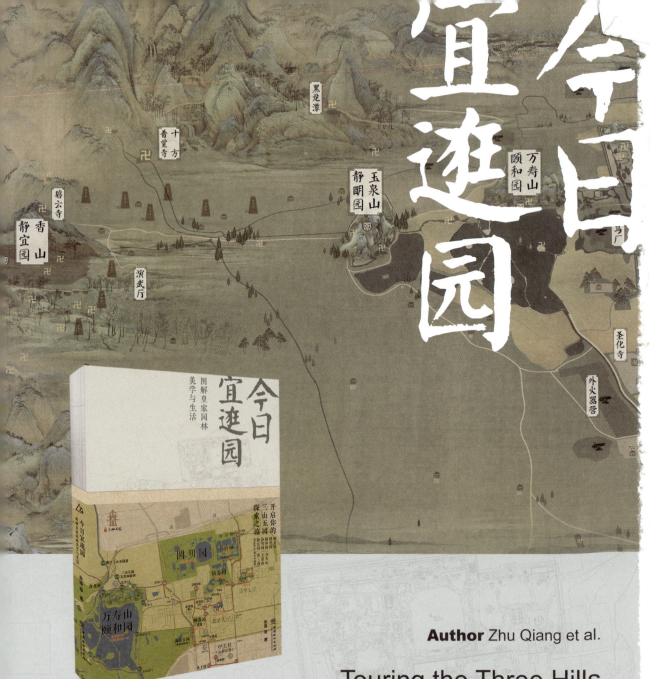

**Author** Zhu Qiang et al.

Touring the Three Hills and Five Gardens

CHINESE EDITION ON SALE

ISBN 978-7-5219-0203-7   |   Price: CNY 98.00